A Colour Handbook on Practical Plant Pathology

nipa books for everyone PAPERBACKS

NIPA® GENX ELECTRONIC RESOURCES & SOLUTIONS P. LTD.

New Delhi-110 034

A Colour Handbook on Practical Plant Pathology

V. K. Yadav
Associate Professor
College of Agriculture
Ganj Basoda, Dist. Vidisha, Madhya Pradesh

nipa books for everyone
PAPERBACKS
NIPA® GENX ELECTRONIC RESOURCES & SOLUTIONS P. LTD.
New Delhi-110 034

nipa books for everyone
PAPERBACKS

NIPA® GENX ELECTRONIC
RESOURCES & SOLUTIONS P. LTD.

101,103, Vikas Surya Plaza, CU Block
L.S.C. Market, Pitam Pura, New Delhi-110 034
Ph : +91 11 27341616, 27341717, 27341718
E-mail: newindiapublishingagency@gmail.com
www: www.nipabooks.com

For customer assistance, please contact
Phone: + 91-11-27 34 17 17
Fax: + 91-11- 27 34 16 16
E-Mail: feedbacks@nipabooks.com

ISBN 978-81-96046-2-79

Composed & Designed by NIPA

Preface

This Laboratory Manual has been designed for students for easy understanding of basic plant pathological laboratory techniques related with Isolation of pathogen, Preservation of disease samples, Demonstration of Koch's postulates, Study of different groups of fungicides and antibiotics, Preparation of fungicides, Methods of application of fungicides, Bio-assay of fungicides, Bio control of plant pathogens and Identification of some important fungal pathogens. It will be useful for students of Plant Pathology and Agricultural Microbiology.

In general a number of constraints often combine to reduce crop production and it is necessary to quantify all of the main constraints. As an assumption about 85 percent of plant diseases are caused by fungi and they are ever-present and are significant constraints in reducing plant growth and crop production. Fungi are multicelled and many species of fungi produce spore, which are reproductive structures that help in dispersal survival and its identification. There are several groups of fungi that are important in causing plant diseases in the nursery. In this manual description of some important causal organisms are also given to make identification easier for students. This manual is designed to help plant pathologists and to enhance learning capacity of students in significant ways. Hopefully, the technical information on fundamental concepts and fungal identification contained herein will certainly useful to the students, scientists and research workers and others, particularly in those areas where access to expert help and advanced facilities is limited. The accurate diagnosis of the cause of a disease helps researchers to recommend successful control measures.

In the past, probably every plant pathology instructor has felt limited supply of images of diseases and pathogens, especially for the introductory courses. Now extensive collections of digitized images are available with each instructor for instructional purposes. Keys used by professionals to identify diseases and pathogens can be modified to include diagrams and photographs along with explanations of new terminologies to enhance students understanding about the subject. Without these enhancements, many key are unusable except by experienced pathologists, who are already familiar with the organisms and their features. Keeping in view all the above points, this manual was prepared with illustrations and it is expected that beginners, students, instructors, scientists and research workers will be benefited from this manual.

V.K. Yadav

Contents

1

Acquaintance with Different Plant Pathological Appliances in the Laboratory

The students in batches will visit the laboratory of Plant Pathology for their acquaintance with different appliances, tools, glassware's and other miscellaneous items, which they will use for various exercises and experiments to be conducted. A list of some of them is given below:

Laboratory appliances/tools

(i)	Autoclave	(ii)	Hot-air oven
(iii)	Refrigerator	(iv)	Incubator
(v)	Hot plate	(vi)	Inoculating needle
(vii)	Knife/Blade	(viii)	Laminar air flow
(ix)	Pan	(x)	Scalpel
(xi)	Scissors	(xii)	Needle
(xiii)	Cork-borer	(xiv)	Bearing blander
(xv)	Spirit lamp	(xvi)	Gel electrophoresis
(xvii)	Forceps	(xviii)	Glass marker
(ixx)	Rotary shaker	(xx)	Centrifuge

Glasswares

(i) Conical flask
(ii) Beaker
(iii) Measuring cylinder
(iv) Pippet
(v) Petri-dishes
(vi) Slides
(vii) Cover-slip
(viii) Watch glass
(ix) Culture tubes
(x) Dropping bottle
(xi) Nematode counting dish
(xii) Baermann funnel

Miscellaneous items

(i) Cotton
(ii) Thread
(iii) Aluminium foil
(iv) Rubber bands
(v) Trays
(vi) Washing brush
(vii) Sieves
(viii) Washing powder
(ix) Blotting paper
(x) Wire basket
(xi) Wash bottle
(xii) Mortar and pestle

2

Preparation of Culture Media for Fungi and Bacteria

Preparation of culture media for fungi (Potato Dextrose Agar medium)

Materials required: Potato peeler, Knife, Muslin cloth, Water, Measuring cylinder, Nonabsorbent cotton, Heating arrangement.

Peeled potato tubers	:	200 g
Dextrose	:	20 g
Agar- Agar	:	20 g
Distilled water	:	1000 ml

Procedure

1. Peeled Potato slices are cooked in 500 ml of water by the time till they are easily penetrated by a glass rod.
2. Filtered with the help of muslin cloth.
3. Agar-agar is melted in another 500 ml of water.
4. Potato extract is added to the melted agar.
5. Dextrose is added in this mixture and stir well with the help of glass rod.
6. Make the volume to 1000 ml, by adding required water.
7. Medium is sterilized in an autoclave at 15 lbs pressure for 20 minutes at temperature of 121.6°C.

Precautions

1. Glassware must be cleaned before using.
2. Rotted potato should not be used for preparation of medium.
3. Do not pour medium over 2/3 capacity of the container for autoclaving.
4. Pour medium quickly to avoid contamination by air spores and close the lid down as soon as possible.
5. Perform the pouring of medium in a cabinet fitted with UV tubes or room with filtered air.
6. Pouring is to be performed near the flame or under aseptic conditions.
7. Medium containing slants or deep tubes are to be stored always at low temperature.

Preparation of culture media for bacteria (Nutrient broth)

Peptone	:	5 gm
Beef extract	:	3 gm
Distilled water	:	1000 ml
pH	:	7

Materials required: Autoclave, pH meter, beaker, measuring cylinder, cotton, culture tube.

Procedure

- Put the 5 gm of Peptone and 3gm Beef extract in 500 ml Distilled water.
- Heat with agitation to dissolve the constituents.
- Make up the volume 1 lit. by adding more water.
- Adjust pH of the medium to 7 by using a pH meter.
- Pour 10 ml per tube.
- Plug the tubes with cotton containing medium.
- Autoclave at 121.6°C, 15 lbs pressure for 15 minutes.
- Allow the autoclave to cool.
- Remove the nutrient broth tubes and store the medium at low temperature in dust free environment.

3

Isolation Techniques

Isolation of fungal pathogens

a. From leaves

- Choose leaves with young lesions.
- The lesions along with some healthy portion are cut in small pieces.
- The leaf portion is surface sterilized with 0.1% $HgCl_2$ for 10-15 seconds and washed with sterilized water 4-6 times.
- After washing, infected leaf pieces are placed on PDA plates with a sterilized forcep.
- After two days of incubation at optimum temperature, bits of mycelium from the margins of leaf pieces are transferred to fresh PDA slants.
- In certain cases the leaves are incubated in moist chamber and growth from infected portion is transferred directly to culture media.

b. From stem

- Split the stem longitudinally from healthy to diseased area.
- Small pieces of diseased tissue are excised from lesion margin.
- Sterilized for 1 – 3 min. in 10% NaOCl and washed in sterile water.
- Transfer 3 – 5 mm tissue onto agar surface.

C. From roots

- Wash the sample to remove excess soil.

- Place in flat-based white bowl containing approximately 2 -3 cm water.
- Carefully tease out roots using scalpel or forceps.
- Detect diseased root portions.
- Cut out leading edges from lesions and sterilize it.
- Place 3-4 pieces per PDA plates.

d. From fruits

- Fruit is wiped with a cotton swab dipped in the disinfectant.
- Fruit is then allowed to dry.
- Using flamed forceps, small pieces of tissue are dig out at margin of infected tissue.
- Place 3-4 pieces per PDA (supplemented with streptomycin sulphate) plates.
- Incubate the plate at 25°C.

e. From soil

- Collect soil samples at random (minimum five) up to a depth of 15-30 cm.
- Mix them thoroughly, dry in air and make fine powder.
- Place 10 g finely pulverized sample into sterile flask.
- Add 90ml of sterile distilled water and stir it for 20-30 minutes (10^{-1} dilution).
- Withdraw 10 ml suspension while the suspension is in motion and add to 90 ml of sterile water (10^{-2} dilution).
- Repeat the process until the desired dilution (up to 10^{-7}) is obtained.
- Pour 1 ml suspension to sterile plate.
- Add 15ml cooled medium (45°C) to each Petri plate and mix the suspension by gentle rotation.
- After solidification of the media, incubate the plate in an inverted position at 26±1°C for 2-7 days.

Isolation of bacterial pathogens

- Cut a small younger portion of diseased tissue.
- Surface sterilization by dipping it in mercuric chloride (1:1000) for few seconds.
- Pass through 3 changes of sterile water.
- Take 1ml of sterile water in a Petri dish and transfer the surface sterilize lesion in it.
- The lesion is cut into small pieces with sterilized razor blade.
- Allow it to stand for 3-10 minutes to permit the diffusion of bacteria into water.
- Pour sterilized molten nutrient agar medium in sterilized Petri plates.
- Allow the medium to solidify.
- Streak a loopful of the suspension on nutrient agar plates by means of the inoculating needle.
- The streaking may be done by gently stroking the needle loop on the surface of the medium in a zigzag manner.
- Inocubate the plates in an inverted position at 25°C for 36 to 72 hrs.
- Observe the plates for appearance of bacterial colonies.
- Select colonies consistently found in large numbers and subculture (Fig.1) them for future use.
- Purity of culture can be checked by making a dilute suspension of culture in water and streak on the nutrient agar plate. If the culture is pure only one type of colony with original characteristics should develop.

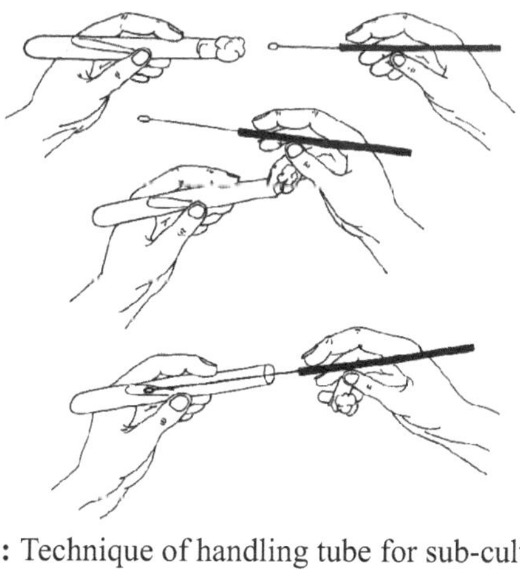

Fig.1: Technique of handling tube for sub-culturing

4

Preservation of Disease Samples

Collection and dry preservation of disease samples

Materials : Iron knife, hand lens, vasculum, pruning scissors, paper envelops and plant press with blotters and news papers.

Method

- Collect infected plant parts and place them in paper envelops (use separate envelops for different kinds of diseased specimens).
- Place envelops in vasculum and bring to the lab.
- Arrange leaves and herbaceous stems inside the folder newspaper sheet.
- Place the newspaper between the blotters in plant press and tighten the bolts.
- After proper drying remove dried specimen from the press and keep them in separate folders.
- Label each folder giving the following informations.

  a. Name of the organism

  b. Habitat

  c. Locality

  d. Date of collection

  e. Name of collector

  f. Collection number

  g. Other notes

The purpose of preservation of plant specimens are

- For herbarium use and storage.
- Dry the material quickly to prevent molding.
- Minimize discolouration.

Wet preservation of plant disease specimens

Materials: Fresh infected leaves, glacial acetic acid, copper acetate crystals, 5% formalin and glass containers.

Procedure: Place leaves in the boiling mixture containing 1% of glacial acetic acid saturated with copper acetate crystals and 4 parts of water. Continue boiling until copper acetate replaces the natural green colour of cells. Take out the material and rinse in water and store in 5% formalin in glass containers. Label the preserved specimen.

5

Demonstration of Koch's Postulates

Koch's Postulates (Pathogenicity test)

Robert Koch developed the following postulates in 1883 on the basis of which an organism could be attributed to be the cause of a specific disease.

i) Recognition: The suspected organism must be found in every case of disease (Presence of organism in the diseased tissues).

ii) Isolation: The organism must be isolated in pure culture from every case of disease (Isolation in pure form).

iii) Inoculation: The isolated pure culture when introduced into the same species or variety of susceptible host must be capable of reproducing the original disease with its typical symptoms (Pathogenicity).

iv) Re-isolation: The organism must be re-isolated from the artificially infected host and its characteristics must be exactly like those observed in step 2 (This step was added by E.F. Smith).

If all the postulates are proved true, then the isolated pathogen is identified as the actual causal organism responsible for the disease.

Precautions to test pathogenicity

- Susceptible plant species are grown under controlled conditions and inoculated with a suspected pathogenic organism.
- When choosing healthy plants to inoculate, it is important to use the same cultivar (variety) from which the pathogen was isolated.

- The symptoms expressed will then be as close as possible to those seen in the original disease (cultivars can differ significantly in susceptibility to a pathogen).
- Selection of techniques in plant pathogenicity test depends upon the pathogen being tested.

Method of stem, foliage and soil inoculation, to test the pathogenicity of a given pathogen

Stem inoculation

1. Pierce the lower stem of the treatment plant with a sterile inoculating needle or hypodermic needle and place a small piece of agar from a pure culture of the pathogen onto the wound site (or inject a small volume of spore suspension into the stem using a hypodermic syringe and needle).
2. Pierce the lower stem of the control plant with the sterile inoculating needle or with the hypodermic needle, but do not treat with the inoculum.
3. Wrap parafilm or plastic wrap over the wounds or injection sites.
4. Water the soil each day.
5. Examine and compare the inoculated plants with the uninoculated plants.
6. Observe and record symptoms and compare these with symptoms observed in the field.

Foliar inoculation

1. Spray the foliage of the treatment plant with a spore suspension (or place a drop of spore suspension on several leaves).
2. Spray the control plant with sterile water (or place drops of sterile water on several leaves).
3. Incubate the pot in a moist chamber or a plastic bag in a greenhouse, avoiding direct sunlight.
4. Examine and compare the inoculated plants with the uninoculated plants.
5. Observe and record symptoms and compare these with symptoms observed in the field.

Soil inoculation

Soil can be inoculated directly using a spore suspension made from a pure agar culture or from a culture grown in flasks. A fungal spore or bacterial suspension can be added post-emergence so that the root system is drenched by the suspension. This method is used as a quick initial test of pathogenicity. A more natural infection process is provided by the admixed or thin layer techniques. Both of these techniques require the production of inoculum on a natural substrate, such as millet seed and rice hulls. Growth of cultures on these substrates in a flask takes 2–3 weeks. A standard amount of inoculum is used for both techniques. Examine and compare the inoculated field with the uninoculated control.

Techniques	Used for
Stem inoculation	*Sclerotinia*, *Sclerotium* and fungal and bacterial wilt pathogens
Foliar inoculation (and moist chamber)	*Septoria*, *Colletotrichum*
Soil inoculation	*Pythium*, *Phytophthora*, *Fusarium*, *Rhizoctonia*
Admixed	*Sclerotium*, *Rhizoctonia*
Thin layer	
Spore suspension (with and without mechanical damage)	*Fusarium* and bacterial wilts

6

Study of Different Groups of Fungicides and Antibiotics

Introduction

Chemicals that kill or retard fungi are fungicides. Most chemicals used to control plant diseases are "protectants" and must be applied before infection to protect the plant from invasion by a pathogen. Protectant fungicides are usually effective against a broad range of fungi and protect the plant against infection on the surfaces of the plant to which they are applied. Often, they require multiple applications during the growing season to maintain coverage as new growth emerges and weathering removes past coverage. A few chemicals are termed "eradicants" since they can eliminate an established infection. Some newer fungicides are systemic, meaning that they are absorbed by the foliage or roots and move within the plant to the site of infection. Systemic chemicals may be protectants, eradicants, or both systemic fungicides can be absorbed by the plant without harming it, and transported to other tissues where they are toxic to fungi. These compounds can control and eradicate established infections, but they are also vulnerable to fungi developing resistance, as they generally only target one step in a biosynthetic pathway to kill the fungus. To minimize the development of resistance by chemical overuse, fungicides are classified into groups based on their chemical activity.

Classification of fungicides on the basis of chemical nature

Sulphur		Copper	Mercurial		Quinone	Heterocyclic nitrogenous compound	Oxathiins	Benzi-midazoles	Others
Inorganic	Organic		Inorganic	Organic					
Powdered sulphur, wettable sulphur, lime sulphur	Ferbam, ziram, thiram, zineb, maneb, nabam, vapam	Bordeaux mixture, Burgundy mixture, cuprous oxide, fixed basic coppers like Copper oxychloride	Mercuric and Mercurous chlorides	Phenyl mercury acetate, methoxy ethyl mercury chloride	Chloranil, Dichlone	Captan, Folpet	Carboxin Oxycarboxin	Benomyl, MBC	PCNB, dodine, tin compounds, antibiotics

Study of antibiotics

Introduction: Antibiotics are one of the most important commercially exploited secondary metabolites. Most of the antibiotic producers used today are the soil microbes. Fungal strains and Streptomyces members are extensively used in industrial antibiotic production. In commercial orchards, antibiotic powders are mixed in a large volume of water to a concentration of 50-200 parts per million (ppm) and applied to tree canopies with a high-pressure sprayer

Definition: Antibiotic is a chemical produced by a microorganism that can, at low concentrations, poison or kill other organisms.

Use: Antibiotics provide a protective barrier on the surface of plants to suppress the growth of the pathogens before infection.

Streptomycin: Streptomycin is an aminoglycoside antibiotic. The primary mechanism of action of streptomycin is binding irreversibly to bacterial ribosomes and thereby inhibiting protein synthesis. At high concentrations, streptomycin can be phytotoxic to plants; thus it is applied to the surface of plants and not injected.

Oxytetracycline: Oxytetracycline is a member of the tetracycline antibiotics. Tetracyclines inhibit protein synthesis by binding reversibly to bacterial ribosomes. This antibiotic is used primarily on peach, nectarine and pear. Tetracycline derivatives are the only antibiotics that are registered for internal use in plants. They are injected into the trunks of palm and elm trees to treat lethal yellows diseases caused by phytoplasmas.

In comparison to most pesticides, streptomycin and oxytetracycline are relatively non-toxic and have been assigned the lowest toxicity rating of the Environmental Protection Agency (EPA). Streptomycin has been the antibiotic of choice as it kills the pathogen (bactericidal). Oxytetracycline, which stops the growth of pathogen but does not kill existing populations (bacteristatic), is less effective than streptomycin (McManus and Jones 1994).

Table: Some of antibiotics produced by BCAs

Antibiotic	Source	Target pathogen	Disease
2, 4 –diacetylphloroglucinol	*Pseudomonas fluorescens* F113	*Pythium* spp	Damping off
Agrocin 84	*Agrobacterium radiobacter*	*Agrobacterium tumefaciens*	Crown gall
Bacillomycin D	*Bacillus subtilis* AU195	*Aspergillus flavus*	Aflatoxin contamination
Bacillomycin, fengycin	*Bacilluss amyloliquefaciens* FZB42	*Fusarium oxysporum*	Wilt
Xanthobaccin A	*Lysobacter* sp. Strain SB-K88	*Aphanomyces cochlioides*	Damping off
Gliotoxin	*Trichoderma virens*	*Rhizoctonia solani*	Root rots
Herbicolin	*Pantoea agglomerans* C9-1	*Erwinia amylovora*	Fire blight
Iturin A	*B. subtilis* QST713	*Botrytis cinerea* and *R. solani*	Damping off
Mycosubtilin	*B. subtilis* BBG100	*Pythium aphanidermatum*	Damping off
Phenazines	*P. fluorescens* 2-79 and 30-84	*Gaeumannomyces graminis* var. *tritici*	Take-all
Pyoluteorin, pyrrolnitrin	*P. fluorescens* Pf-5	*Pythium ultimum* and *R. solani*	Damping off
Pyrrolnitrin, pseudane	*Burkholderia cepacia*	*R. solani* and *Pyricularia oryzae*	Damping off and rice blast

Source: The Plant Health Instructor, 2006

7

Preparation of Fungicides (Bordeaux mixture, Bordeaux paste, Cheshunt compound)

Preparation of Bordeaux mixture

Materials: Iron, beaker, copper sulphate, lime, weighing balance, plastic container, wooden stick and muslin cloth.

Bordeaux mixture derives its name from the place of its discovery, Bordeaux, France.

Ingredients required for the preparation of 100 litre of 1 percent elicoid mixture are:

1. Copper sulphate (blue vitriol or bluestone) : 1 kg
2. Quick lime (unslaked) : 1 kg
3. Water 100 litres

Method

- Dissolve 1kg powdered copper sulphate crystals in 50 litres of water in a mudpot or plastic bucket.
- In another 50 litres of water dissolve 1kg of quicklime.
- Stir thoroughly with the help of wooden stick.
- Put a filter cloth on the mouth of 3rd container.
- Pour the contents of both the containers in a 3rd container simultaneously and stir it thoroughly.

- Make up the desire volume (ie. 100 litres).
- Test the mixture before use for the presence of free copper, which is harmful to the plant.

How to test Bordeaux mixture

a) If the Bordeaux mixture contains excess of lime, it turns the red litmus paper blue and if it contains excess copper, it turns the blue litmus paper red. An excess of copper compound in the mixture is toxic to foliage.

b) Dip a knife or iron blade or nail in the Bordeaux mixture and observe if copper deposits on it. A deposit of copper on the metal indicates that the mixture is unsafe. Add more lime in the solution to make the mixture safe.

c) A more accurate test can be made with ferrocyanide of potassium. Add a few drops of the Bordeaux mixture to small quantity of ferrocyanide reagent in a saucer. If no change occurs, the mixture is safe but if it turns into reddish brown colour, it indicates that copper compound is present in excess. So, add more lime solution to make the mixture safe.

Use

- Used against a range of fungal and bacterial diseases such as leaf spots, blights, anthracnose, downy mildew and cankers etc.
- The Bordeaux mixture also acts as pit disinfectant in root/collar rot infested sites before planting new saplings (a young tree) in orchards.
- When used as a basin drench treatment, it controls the spread of root rot pathogens from diseased plants to healthy ones.
- It renders protection to pruning cuts or injuries, when used as a prophylactic treatment.

Merits

1. Its natural adhesiveness or tenacity.
2. Its relative cheapness.
3. Its utility in controlling a wide variety of diseases.
4. Its safety to handle.
5. Can be prepared in remote areas.

Demerits

1. Its phytotoxicity on certain crops i.e. apple, peaches etc.
2. It delays ripening.
3. Botheration of preparing.
4. Corroding action on metallic containers of spraying equipment.

Precautions

1. Bordeaux mixture should not be applied, when it is raining or on exceptional hot day when the plants are showing sign of temporary wilting and it should not be sprayed, particularly on newly emerged tender foliage.
2. Fresh Bordeaux mixture should be sprayed as it loses its fungicidal activity on storage.
3. Do not use metallic containers for preparing copper sulphate solution and lime suspension. Use a wooden stick for stirring to get a homogenous mixture. Never use metallic stick.
4. Do not use the Bordeaux mixture in combination with any other chemical or pesticide e.g. in citrus, zinc sulphate spray should be given, keeping a gap of at least one week from that of the Bordeaux mixture.
5. To avoid choking of the nozzle, it is advisable to strain the Bordeaux mixture through a cloth or a sieve before putting it into the spray tank.
6. The Bordeaux mixture tends to sediment easily. Therefore, its stirring is desirable.
7. It is not advisable to spray the Bordeaux mixture on fruit-laden trees as the spray may inflict russetting, especially on apple and pear.
8. After carrying out spray operations the appliances should be thoroughly washed with plenty of water to remove any copper deposits.
9. The left over Bordeaux mixture should not be dumped in the field as this may prove toxic to the subsequent sowings.

Bordeaux paste (Bordeaux mixture 10%)

Bordeaux paste consists of the same ingredients as that of Bordeaux mixture, but it is in the form of a paste. It is normally prepared by adding

100 gm of copper sulphate and 100 gm of quick lime each in 500 ml of water separately. Mix them together to make one litre of paste.

Use

- The Bordeaux paste is the most effective wound dressing material. The fine layer formed on the cut ends kills the pathogen and protecting it from subsequent invasion and results in rapid healing of the wounds.
- Effectively check the attack of collar rots and root rots of fruit trees.

Cheshunt compound

It contains two part of copper sulphate and 11 part of ammonium carbonate. These two are well powdered and thoroughly mixed and the dry mixture is stored in an air tight receptacle for 24 hrs before using. If left exposed, it gradually losses ammonia and becomes less effective.Thirty grams of this mixture is dissolved in a little hot water and the solution is made up to 9 litres with cold water. This solution is then ready for spraying.

8

Methods of Application of Fungicides

Introduction

Diseases are a common occurrence on plants, often having a significant economic impact on yield and quality, thus managing diseases is an essential component of production for most crops. These fungicides are used because of three main reasons:

a. To control a disease during the establishment and development of a crop.

b. To increase productivity of a crop and to reduce blemishes.

c. To improve the storage life and quality of harvested plants and produce.

Application Methods

To protect the crop from fungal pathogens, fungicides are applied in various formulations (as dust, granules, gas and most commonly, liquid). They are applied to:

1. Seed treatment
2. Soil treatment
3. Protection of standing crop

1. Seed treatment

a. Dry mixing

b. Short wet method

c. Slurry (most effective method)

d. Seed dip method

The seed treatment with fungicide is essential because a large number of disease causing fungi are carried on and/or in the seed and when the seed germinates, these fungi also become active and cause either seed or seedling mortality or produce diseases at later stage. The purpose of seed treatments by the use of fungicides is to destroy seed borne fungi that cause seedling blight, seed decays or other disease such as smuts. Also protect from soil inhabiting fungi.

In fact, treatment of seeds of all crops should be done as routine practice since it is a type of cheap insurance against possible disasters at a later stage. There are various types of seed treatments.

Fungicidal seed treatment involves the use of more than one fungicide since there is no one fungicide that may be recommended as a universal preventive for all of the fungi which can attack seeds and seedlings. Simple seed treaters the most common one available in India is the rotary seed dresser in which dry seed dressing is done. On the other hand, almost all the seed processing plants, owned by the government as well as private seed producers, have slurry treaters. The process of slurry treatment in slurry treaters is a continuous one and is more efficient than the rotary seed dressers.

Other methods of seed treatment are seed dip and planter or hopper box treatment. The seed dip method involves preparing fungicide suspension in water often at field rates and then dipping the seed or propagative material for special time.

2. Soil treatment

Many diseases arise from the soil and that if steps are taken to eliminate the inoculum or reduce the inoculum potential, one hope to achieve disease control.

The soil treatment methods involving the use of chemicals are:

a. Soil drenching: Fungicides solutions are made up with water at about the same concentration as for spraying and applied to the soil surface either before or after plants emerge. Fungicide in general reaches a depth of at least 10-15 cm. This method is followed for controlling damping off, root rots or infections at the ground level where per unit area discharge high. Drenching is used for precious crop, nursery bed, kitchen garden purpose to control disease in row and furrow.

b. Broadcast: Sometimes non volatile fungicides mixed with soil or fertilizer are scattered with hand as uniformly as possible over the field and then mixed with the soil with a suitable implement. This method involves consumption of greater quantity of fungicides than other methods and is followed where suitable implement for proper application of fungicide is not available.

c. Furrow application: In this method fungicides are applied either as dust or with water to the furrow at the planting time. Such a treatment is for the control of diseases that occur at the base of the plant. This method requires much less quantity of fungicides per hectare than the broadcast method.

d. Fumigation: Application of certain chemicals to the soil can control fungi and nematodes. Such materials usually produce gas that distributes itself through the soil. Some impervious cover, like thin polythene sheet, which confines the gas to the soil for a limited period of time is required, this method is usually restricted to small areas and high value crops.

3. Protection of standing crop

a. Spraying: Wettable powders are most commonly used for preparing sprays. The most common diluents or carrier is water. The dispersion of the spray is usually achieved by its passage under pressure through nozzles of sprayers.

   Spraying is of three types, viz. high volume, low volume, ultra low volume. When spray involves large quantities of liquid per unit area, they are termed high volume generally. 0.1 to 0.3% suspension of fungicide is prepared for foliar application.

   High volume: Six hundred litres water used, droplet size larger 50-100 micron. eg. Knap sac sprayer.

   Low volume: 50-100 litre water used, droplet size 10-15 micron eg. Power sprayer.

   Ultra low volume: Instead of suspension slurry used, droplet size extremely minute. eg. Jets, aeroplane, helicopter used.

b. `Dusting: Dusts are applied to leaves, fruits and stems of plants as an alternative to spraying. Dry powders are used for covering host surface. In these dust the active component is diluted with a suitable finely divided "carrier". Generally dusting is practicable only in calmest weather and a better protective action is obtained if the dust is applied when the plant is wet with dew or rain.

c. Fungicide pastes or paints: Fungicides are occasionally mixed with water, alcohol or other carriers and applied as paint on a wound surface. For example Bordeaux mixture can be made up to paint like consistency by mixing a prepared Bordeaux mixture powder with linseed oil or water for use in painting tree wounds.

9

Bio-assay of Fungicides

Poisoned food technique

The principle involved is to poison the medium with fungi toxicants and then allow a test fungus to grow on such medium. The test may be performed both on solid and liquid medium.

Procedure

On solid medium

- Prepare potato dextrose agar medium.
- Add required amount of fungicide and thoroughly mixed.
- The medium is poured into sterilized Petri-plates.
- A culture of test fungus is grown on PDA for a certain period (generally seven days) at the optimum temperature for growth.
- Small disc (7 mm) of fungus culture is cut with sterile cork borer and each Petri plates are centrally inoculated aseptically containing the medium with desired fungicidal concentration.
- Suitable checks are kept where the culture discs are grown under the same conditions on PDA without fungicide.
- The fungal colony diameter is measured at every 24 hr.

The radial growth of the colony was recorded when maximum growth was observed in control and it was calculated by using the following formula:

$$I = \frac{C - T}{C} X\ 100$$

Where,

I = Per cent inhibition

C = Radial growth in control

T = Radial growth in treatment

On liquid media

- Any fungus which can grow well on a liquid medium can be used as test fungus.
- Liquid medium such as Blakeslees medium or Richards medium have been commonly used.
- Prepare the desired liquid medium, and pour in measured quantity in Erlenmeyers flasks.
- In luke warm medium, required quantity of fungicidal dilutions are added by pipette.
- Add fungal agar disc to this medium.
- Control flasks would contain the same except the fungicide solutions.
- After incubating the flasks at suitable conditions either the rating is done on 1-10 scale or actual dry weight of the fungus is taken after filtering the liquid out.

Inhibition zone technique

Preparation of assay plates

- Fifteen ml of medium is poured in each of the sterilized Petri dish and let it be solidified.
- Five ml of warm seeded agar is applied at the surface of the medium on each plate (warm seeded agar is prepared by cooling the molten agar to 40°C and then adding the amount of spore suspension [(0.5-1X10^6) from viable test fungal spores/ml of seeded agar)].

- While the seeded agar is still molten, the plate is tilted to ensure even coverage before the agar solidifies.
- These covered Petri dishes are stacked in a refrigerator (2-4°C) upside down to prevent condensation of moisture.

Test fungus

- Any fungus which sporulates easily can be used as the test fungus for following this technique.

Test sample

- A series of dilutions or the test chemicals are prepared and expressed as ppm or as µg/ml.

Setting up the assay

- Seven filter paper discs of 12.7 mm diameter (evenly spaced) are spread individually (one in centre and six in outer portion) with a needle and planted upon the chilled seeded agar of the plates. The liquid should be applied immediately after the pads are planted upon the agar. Volume of the test liquid should be slightly less than required to saturate pad. For a 12.7 mm disc, 0.09 ml applied with a 0.1 ml micropipette is found satisfactory.

Incubation

- An incubation period of 16-30 hr is given at 28-30°C depending upon the test fungus before the clear zones of inhibition around each disc are measured from the centre of the paper disc.

Estimation of toxicity

- For estimating relative toxicity to know the efficacy of the material being evaluated, a reference fungicide of known fungitoxicity is included in the test. Material with low diffusibility can not be properly evaluated by this technique.
- The principle of inhibition zone technique has been also followed in determining the fungicide quantities on the treated seed to determine the efficacy of seed treatment.
- In order to find out the quantity of fungicide present on treated seeds, the zones of inhibition obtained are compared with the standard curve and the amount calculated.

Slide germination method

- Slide germination method was given by Reddick and Wallace (1910).
- Spray the glass slide with pure concentration of fungicidal suspension prepared in sterile distilled water. (The deposition of test chemical is calculated by weighing the slides before and after application of fungicide)
- Allow it to dry.
- Culture the fungus on suitable medium.
- Prepare the spore suspension in distilled water (500 spores/ml for large spores and 50000 spores/ml for small spores).
- One drop (0.05 ml) of spore suspension is placed on the treated dry slide.
- Incubate the slides at suitable temperature in a moist chamber.
- Germination percentage is recorded after 16 to 20 hrs.

***Note*:** *Spore is considered germinated, if the spore germ tube length is equal to 1-1.5X the length of the spore.*

10

Bio-control of Plant Pathogens

Biological control of disease employs natural enemies of pests or pathogens to eradicate or control their population. Interactions that lead to biocontrol can include antibiosis, competition, induction of host resistance, and predation. Some of the microbial taxa that have been successfully commercialized and are currently marketed include bacteria belonging to the genera *Agrobacterium, Bacillus*, *Pseudomonas* and *Streptomyces* and fungi belonging to the genera *Ampelomyces, Candida, Coniothyrium* and *Trichoderma.*

- Dual Culture Technique: For in vitro screening, performance of antagonist isolates in dual culture (Morton and Stroube, 1955) with test pathogen can be tested.
- Pour 15-20 ml of sterilized melted PDA in 9 cm Petriplates and allowed to solidify.
- Place the bits (5 mm) of the test pathogen and antagonist on the PDA plates opposite to each other from 1.0 cm from the periphery of plates (if both are fast growing) or place it 2-3 cm apart (if both are slow growing).
- The Petriplates are incubated (at 28 ± 2°C) for desired duration.
- Periodical observation on the growth of the antagonists and the ability of the antagonist to colonise the pathogen are recorded.

Seed treatment: Diseases of vegetable and field crops often are caused by fungi or bacteria carried on, in, or with the seed. Disinfecting seed to kill these potential pathogens is important in preventing introduction of pathogens into a production area as well as protecting crops in the current cycle from disease losses. Addition of biological control agents onto the seed surface or seed bed control seed pests by parasitizing the pest organisms, competing for food on the root system, or producing toxic compounds that inhibit pathogen growth and protect seeds as well as developing seedlings from decay or damping-off caused by soil borne organisms.

Precaution and method of treating seed with biological control agents

- Most biological control agents are sensitive to environmental conditions during storage and must be handled with care. Avoid storage at hot temperatures. Follow manufacturer's label directions for handling biological agent and treated seeds.
- For seed treatment shake seed and biological agent together 3 to 5 minutes in jars or cans (not over half full) with tight-fitting lids. For small packets, add a pinch of dust to seed in the packet and shake well. For larger quantities of seed, use a rotary or mechanical mixer.

11

Study of Some Important Fungal Pathogens

Some basic informations about genera of Family Pythiaceae, Peronosporaceae and Albuginaceae;

Pythiaceae

Indeterminate (grows continuously) sporangiophores, have sporangia that are borne variously, but not in chains, on somatic hyphae or on sporangiophores. This family includes the genera *Pythium* and *Phytophthora*.

Peronosporaceae

Determinate sporangiophores, form sporangia on well-defined, branched sporangiophores of determinate growth. There are seven genera in the family including *Basidiophora, Bremia, Bremiella, Peronospora, Plasmopara, Pseudoperonospora* and *Sclerospora*.

Albuginaceae

Club shaped sporangiophores, catenate sporangia, produces chains of sporangia on club-shaped sporangiophores of indeterminate growth. The Albuginaceae contains a single genus *Albugo*.

Organisms in the Oomycota were once classified in the kingdom fungi, but are now included in the kingdom stramenopila (Chromista).

Hyphal walls of the oomycetes contain cellulose. The group is characterized by production of oospores. Oospore wall consist of three parts:

- Exospore
- Epispore
- Endospore

Second reproduction is by gametangial contact between antheridia and oogonia.

Hypogynous : Antheridium formed in oogonial stalk.

Paragynous : Antheridium attaches laterally to the oogonium

a) Monoclinous: branch off oogonial stalk.

b) Diclinous: branch off separate hyphae.

Amphigynous: Oogonium grows through Antheridium

11.1 Study of *Pythium, Phytophthora* and *Albugo*

Pythium

Mycelium: Coenocytic, hyaline

Sporangiophore: Indeterminate

Sporangia: Sporangia may be indistinguishable from somatic hyphae or quite distinct, depending on the species. At the time of germination a bladder like vesicle is formed at the apex and protoplasm flows into it. The contents of sporangia migrate into vesicles where zoospores are differentiated.

Asexual spore: In the vesicle, 100 or more zoospores are produced and germinate by producing a germ tube.

Sexual structures: Oogonium (female reproductive organ): spherical, mostly terminal, rarely intercalary.

Antheridia (male reproductive organ)**:** club shaped, either terminal or intercalary.

Resting structure: Oospore thick walled structure resistant to adverse conditions and serve as survival of the fungus.

Diseases: Damping off of Papaya (*Pythium sp.*), Damping off of tomato (*P. aphanidermatum*), Collar rot of Papaya (*P. aphanidermatum*), Pythium blight of turf grass (*P. aphanidermatum* and *P. ultimum*), Cucumber and sugar beet damping-off (*P. ultimum)*, Root and stem rot of chrysanthemum (*P. dissotocum*, *P. oedochilum*, *P. sylvaticum*, *P. ultimum* var. *ultimum* and *P. elicoids),* Damping-off of cucumber (*P. ultimum*).

Other facts

- The germination of sporangia and oospore is influence by temperature. Temperature above 18°C favour germination by germ tubes, whereas temperature between 10-18°C induce germination by zoospores. Some of the more important *Pythium* species include *P. ultimum, P. aphanidermatum,* and *P. debaryanum*.
- The most notable difference between the two genera (*Phytophthora* and *Pythium*) is that zoospores of *Phytophthora* species are delimited and functional within the sporangium, whereas in *Pythium* species, zoospores are formed from the cytoplasm of sporangia that have migrated into a vesicle.

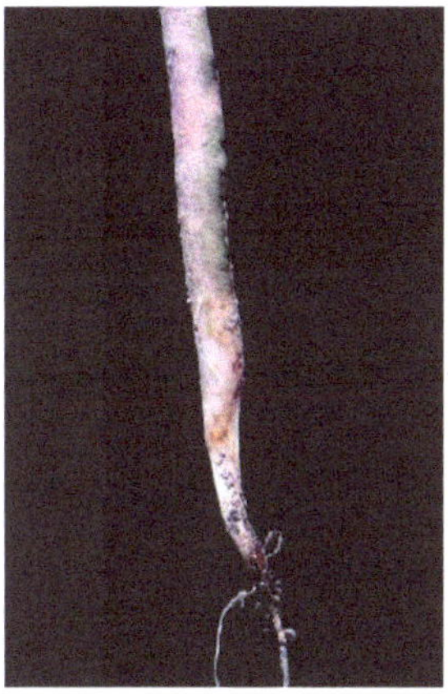

Damping off

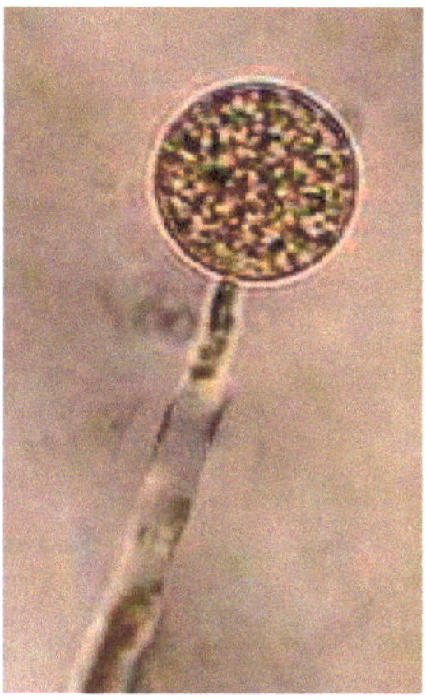

Sporangia

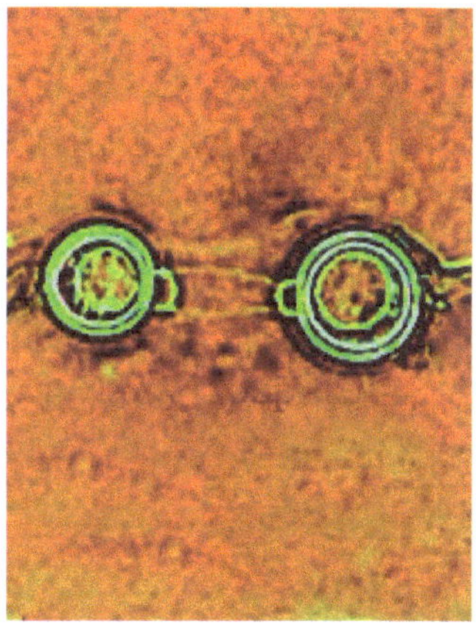

Oospore

Phytophthora

Mycelium: Coenocytic, hyaline, much branched.

Sporangiophore: Slender, hyaline, branched, indeterminate, sympodially branched.

Sporangia: Sporangia are thin walled, hyaline, oval or pear shaped, papillate, semi papillate or non papillate.

Asexual spore: Zoospores are biflagellate, tinsel (posses fine hairs) and whiplash (without hairs).

Sexual structures: Oogonia (female reproductive organ): Thick walled, become rough and yellowish brown with age.

Antheridia (male reproductive organ): spherical or oval.

Resting structure: Oospore: Germinate by germ tube.

Chlamydospores: Thick walled, may be abundant few or absent according to species.

Diseases: Leaf blight of clocasia (*Phytophthora colocasiae*), Late blight of potato and tomato (*Phytophthora infestans*), Red stele of strawberry (*P. fragariae* var. *fragariae*), Root and stem rot of soybean (*P. sojae*), Blackshank of tobacco (*P. nicotianae*), Foot rot of citrus (*P. nicotianae*), Root rots of forest and nursery crops and avocado (*P. cinnamomi*), Sudden oak death (*P. ramorum*), Oak death (*P. quercina*), Black pod of cacao (*P. megakarya*), Fruit rot in coconuts and betel nuts (*P. palmivora*), Cinnamon root rot (*P. cinnamomi*), Alder root rot (*P. alni*), Rhododendron root rot (*P. cactorum*) etc.

Other facts

Formation of chlamydospores can also be stimulated by poor aeration and high carbon dioxide concentration in the soil atmosphere.

Leaf blight of clocasia

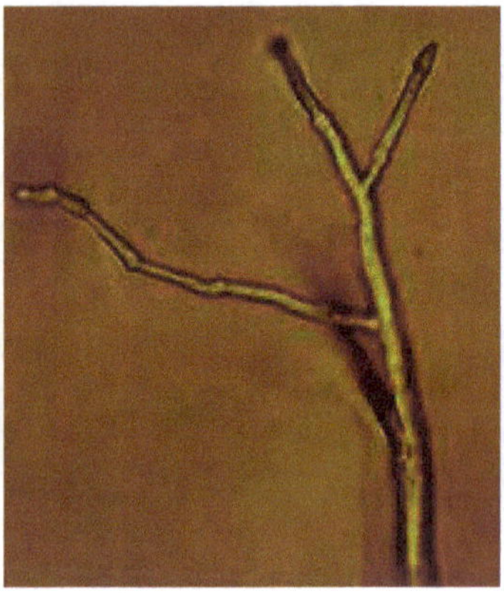

Sporangiophore

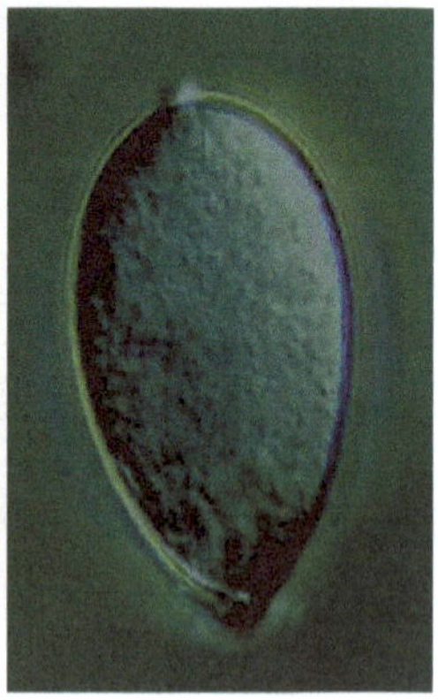

Sporangia

Albugo

Mycelium: Coenocytic, hyaline, intercellular, with knob shaped haustoria.

Sporangiophore: Club-shaped sporangiophores formed under the epidermis, free from each other laterally and are very thick, especially at the base.

Sporangia: Hyaline, nearly spherical, formed in basipetal succession in chains. The first formed being thick walled and at the top and is incapable of germination. Its only function appears to be to raise the epidermis and finally rupture it thus facilitating the dispersal of the rest of the sporangia. The germination of sporangia occurs readily when they are completely immersed in water. Sporangia germinate by germ tube or by formation of zoospores.

Asexual spore: Zoospore: Four to eight numbers in each sporangium. Optimum temperature for their germination is about 10°C at which the rate of germination and the number of zoospores formed are maximum.

Sexual structure: Oogonia (female reproductive organ): Globose, terminal or intercalary, consist of 100 or more nuclei.

Antheridia (male reproductive organ): Clavate, paragynous and contains 6-12 nuclei.

White rust of crucifers

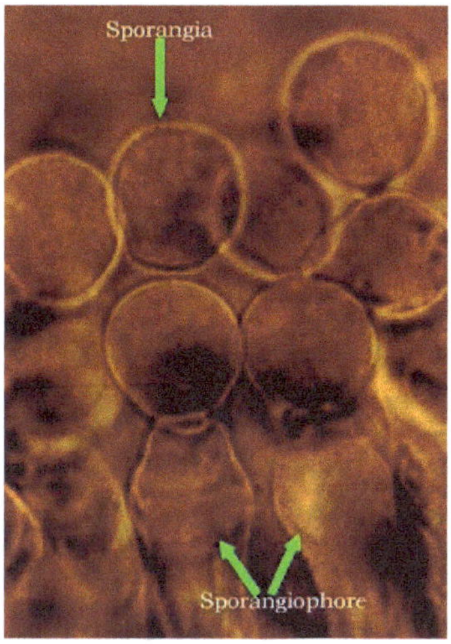

Sporangiophore and sporangia of *Albugo*

Resting structure: Oospore: Resting period of several months, globose, chocolate olive brown, terminal or intercalary, 30-35µm.

Diseases: White rust of crucifers (*Albugo candida*), White rust of Amaranth (*A. bliti*), White rust of Arabidopsis (*A. laibachii*), White rust of Spinach (*A. occidentalis*), White rust of Tragopogon (*A. tragopogonis*), White rust parasitic to some Convolvulaceae (*A. ipomoeae-panduratae*).

11.2 Study of *Sclerospora*, *Peronosclerospora*, *Pseudoperonospora*, *Peronospora*, *Plasmopara* and *Bremia*

- Organisms in the Oomycota were once classified in the kingdom Fungi, but are now included in the kingdom Stramenopila (Chromista).
- Sexual reproduction is by gametangial contact between oogonia and antheridia (meiosis occurs in the gametangia).
- Hyphae are coenocytic, contain cellulose in the wall, and have diploid (2N) nuclei.

- Some downy mildew genera have a more restricted host range, e.g. *Basidiophora, Paraperonospora, Protobremia* and *Bremia* on Asteraceae; *Perofascia* and *Hyaloperonospora* almost only on Brassicaceae; *Viennotia, Graminivora, Poakatesthia, Sclerospora* and *Peronosclerospora* on Poaceae, *Plasmoverna* on Ranunculaceae. The largest genera, *Peronospora* and *Plasmopara*, have a very wide host range. The genera are distinguished by the highly differentiated branching patterns of sporangiophores that always exhibit determinate growth.

Sclerospora

Mycelium: Coenocytic, intercellular in the parenchymatous tissues.

Sporangiophore: Long, bulbous, stout, hyaline, repeatedly dichotomously branched on the upper part, arise through the stomata, many upright, short terminal branches near the end, branch irregularly.

Sporangia (spore-containing "sacks"): Hyaline, thin walled, elliptical, sporangia are produced during the night under moderate temperatures and high humidity. Optimum sporangium production occurs at 20°C.

Zoospores: Sporangia germinate to liberate 1 to 12 zoospores, which encyst and germinate by germ tube.

Oogonia: Globose, female reproductive structure.

Antheridia: Clavate, male reproductive structure.

Resting structure: Oospores are thick-walled, spherical, brownish yellow and 22 to 35 µm in diameter.

Diseases: Downy mildew and Green ear disease of bajra (*S. graminicola*), Downy mildew of sugarcane (*S. sacchari* and *S. spontanea*), Downy mildew of maize (*S. maydis* and *S. philippinensis*), Downy mildew of sorghum (*S. sorghi*), Downy mildew of grasses (*S. graminicola* and *S. philippinensis*), Downy mildew of sorghum (*S. sorghi*), Downy mildew of grasses (*S. graminicola*).

Green ear disease of bajra

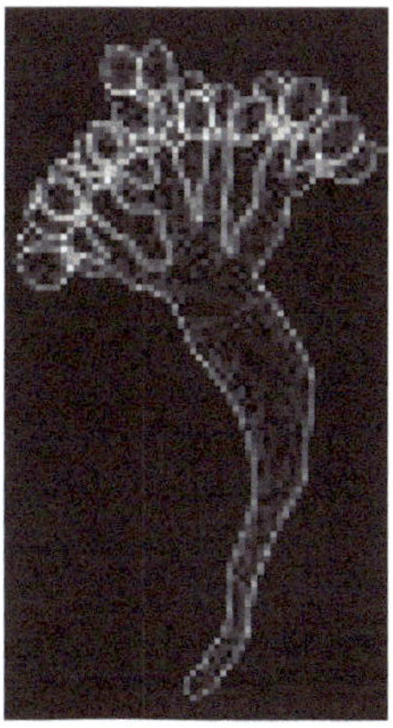

Sporangiophore & Sporangia of *Sclerospora*

Peronosclerospora

Mycelium: Coenocytic.

Sporangiophore: Erect, spreading, fragile, hyaline, long, usually dichotomously branched, emerge singly or in groups from stomata on the lower and sometimes the upper sides of the leaves.

Sporangia: Hyaline, oval to subspherical and are borne on elongated, tapered sterigmata.

Oogonium: Spherical, reddish brown, irregularly elliptical female reproductive organ.

Antheridia: Male reproductive structure.

Resting structure: Oospore: Hyaline and spherical with light yellow outer walls.

Diseases: Downy mildew of Sorghum and maize (*Peronosclerospora sorghi),* Downy mildew of maize (*P. maydis* and *P. philippinensis*), Downy mildew of sugarcane and maize (*P. sacchari*).

Other facts

The species of *Sclerospora* that showed conidial germination of sporangia and no zoospore production, were transferred to this genus.

Downy mildew of maize

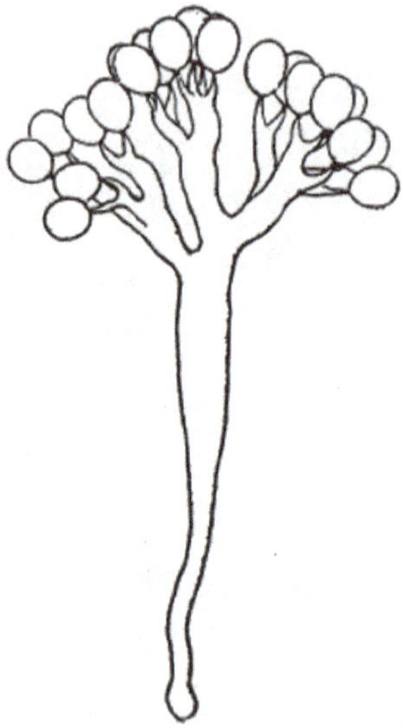

Sporangiophore & *Sporangia* of *Sclerospora*

Pseudoperonospora

Mycelium: Coenocytic, intercellular with small, ovate haustoria.

Sporangiophore: Branched dichotomous or trichotomous, sterigmata sub-acute.

Sporangia: Grayish to purplish, ovoid to elliptical, thin walled and papilla at the distal end. Sporangia germinate by zoospores.

Zoospore: Biflagellate

Antheridia and oogonia, which on fertilization produce the oospore (resting structure).

Oospore: Spherical, rarely obovid to ellipsoid, light yellow and smooth walled.

Diseases: Downy mildew of cucurbits (*Pseudoperonospora cubensis*), Downy mildew of hops (*P. humuli*).

Downy mildew of cucurbits

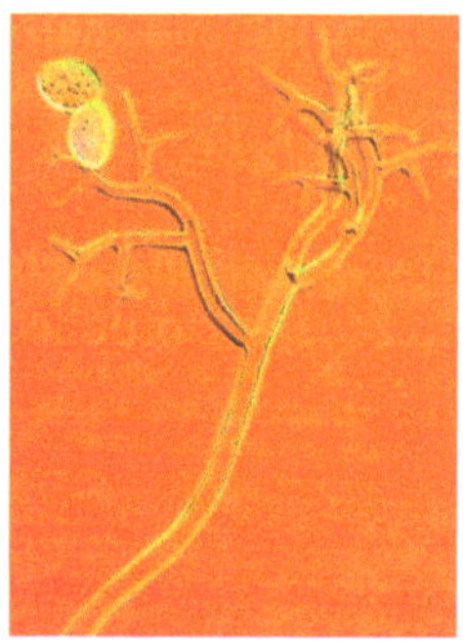

Sporangiophore & *sporangia* of *Pseudoperonospora*

Peronospora

Mycelium: Coenocytic, intercellular.

Sporangiophore: Non septate, dichotomously branched at acute angles, 6-8 times at the tip, emerge singly or in groups through stomata on the undersurface of the leaf, unbranched for a major portion of their length, sterigmata long, curved, slender, pointed.

Sporangia: Single conidia borne at the tip of each branch, broadly oval, ellipsoidal, hyaline, germinate by lateral germ tube.

Resting structure: **Oospore**: Surrounded by thick smooth epispore wall, these are globous, germinate by germ tube but in *P. tabaciana* by means of zoospore.

Diseases: Tobacco blue mold (*Peronospora tabacina*), Downy mildew of Onion (*P. destructor*), Downy mildew of soybean (*P. manchurica*), Downy mildew of Snapdragon (*P. antirrhini)*, Downy mildew of pea (*P. pisi*), Downy mildew of Lucerne (*P. viciae*), Downy mildew of spinach (*P. effusa*), Downy mildew of beet (*P. farinosa*), Downy mildew of clover and alfalfa (*P. trifoliorum*).

Downy mildew of Onion

Sporangiophores & sporangia of Peronospora

Plasmopara

Mycelium: Coenocytic, hyaline, intercellular.

Sporangiophore: Stiff, erect, short terminal branches, 4-6 branches at nearly right angles to the main stem of the sporangiophore and each branches produces 2, sometimes 3, secondary branches in similar manner. Secondary and tertiary branches on the sporangiophore occur at right angles to the axis and the sporangia are borne on their blunt tips. The tips of sporangiophores are somewhat swollen with peg-like appendages. Sporangiophores emerge on the underside of the leaves and on the stems through stomata or rarely by pushing directly through the epidermis. In the young fruits, sporangiophores emerge through lenticels.

Sporangia: At the tip of the secondary branches, single, lemon shaped sporangia are produced.

Asexual spore: Germination of sporangia is either by production of zoospores (1-6) or through germ tube depending upon humidity. Zoospores are pear shaped, have two apical flagella.

Sexual structure: Antheridia and oogonia, which on fertilization produce the oospores, mostly inside the host.

Diseases: Downy mildew of Grape (*Plasmopara viticola*), Downy mildew of members of umbeliferae (*P. nivea*), Downy mildew of cucurbits (*P. australis*), Downy mildew of Sunflower (*P. halstedii*), Downy mildew of *Geranium pretense* (*P. pussila*).

Downy mildew of grape

Sporangiophore of *Plasmopara*

Bremia

Mycelium: Coenocytic, hyaline, intercellular.

Sporangiophore: Dichotomously branched, tips of the branches are elongated to form a saucer shaped structure bearing a ring of 3-6 sterigmata and these bear papillate sporangia at their tips. Arise singly or in groups through stomata.

Sporangia: Ovate, papillate and mostly sporangia germinate by germ tube.

Asexual spore: Zoospore: Hyaline, globular, biflagellate.

Resting structure: Oospore: Spherical, rarely elliptical, aplerotic, smooth walled.

Diseases: Downy mildew of Lettuce (*Bremia lactucae*).

Downy mildew of Lettuce

Sporangiophore of *Bremia*

11.3 Study of genera *Mucor* and *Rhizopus*

Mucor

Mycelium: Coarse, coenocytic and branched and usually taper to fine points.

Sporangiophores: Hyaline, have various branches arising from the main branches and do not produce rhizoids at their bases.

Sporangium: Each sporangiophores bears a globose or spherical multispored columellate sporangium at its tip. In some species the immature sporangium may be circinate, but it eventually assumes an upright position.

Columella: The columella is enlarged at the basal region of the sporangium, which is delimited by a wall from the spore filled area. At maturity when the sporangial wall deliquesce, the spores adhere to each other, forming a tight mass which remains firmly attached to the columella and the spores are, therefore not blown off by wind. They are disseminated by splashing raindrops, insects and ants. Species of mucor, therefore, do not cause contamination of laboratory cultures.

Zygospores: The brown or black thick walled zygospores are formed in the aerial mycelium and are covered with rough blunt conical projections or warts.

Diseases: Decay of pome and stone fruits (*M. piriformis*).

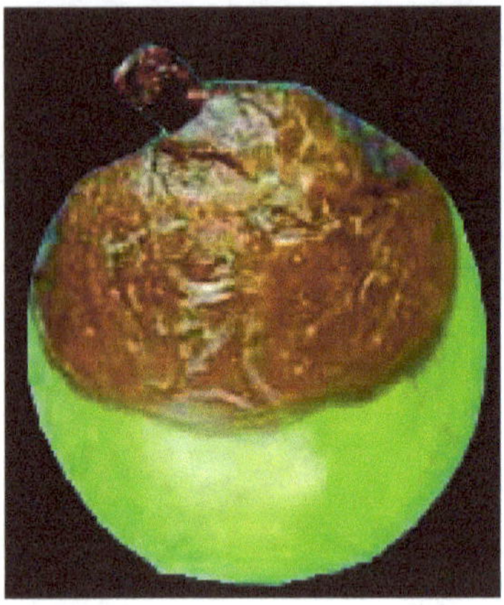

Decay of stone fruit

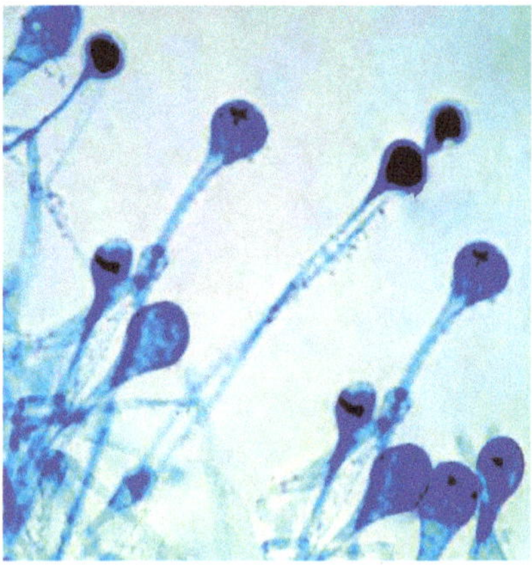

Sporangiophores of *Mucor*

Rhizopus

Mycelium: Coenocytic, which has prominent aerial branches known as stolons.

Stolons: Differentiation of the mycelium into horizontal aerial hyphae called stolons, the mycelium produces many aerial stolons that develop rhizoids at certain points. When stolons come in contact with the substratum the rhizoidal branches act as absorptive organs. It is the chief features of *Rhizopus* that distinguish it from *Mucor.*

Head rot of sunflower

Sporangiophore of *Rhizopus*

Sporangiophores: Directly above the rhizoids, one or more sporangiophores are produced. They are globose to oval and multinucleate, usually unbranched, grow in clusters (umbels). The top of each sporangiophores become swollen as it reaches maturity and sporangium begins to develop. Spores germinate by a germ tube that develops into a fluffy, much branched, white, aerial mycelium.

Diseases: Soft rot of vegetables (*Rhizopus* spp.), Head rot of sunflower (*R. arrhizus, R. stolonifer* and *R. microsporus*), Post harvest rotting of stone fruits (*R. stolonifer*), Whisker's or storage rot of apple (*R. arrhizus*), Leaf spot of jack fruit (*R. atrocarpi*), Black mould on stone fruits (*R. nigricans*).

Other facts

- Rhizopus (Gr. Rhiza=root+pous=foot)
- Most species are heterothallic but *R. sexualis* is homothallic. For heterothallic species of *Rhizopus*, sexual reproduction requires the presence of two physiologically distinct (+ and -) compatible mycelia.
- Rhizoids are not produced in *R. arrhizus.*
- *Rhizopus* are used in the manufacture of lactic acid and an important source of fumaric acid.

11.4 Study of *Oidium, Oidiopsis, Ovulariopsis, Erysiphe, Phyllactinia, Uncinula* and *Podosphaera*

Three types of conidiophore can be recognized in the family Erysiphaceae.

1. The oidium type, which consist of a short stipe of one or more cells, a germinative or spore mother cell which give rise to a chain of maturing conidia. This is most common particularly in species of *Erysiphe.*
2. The ovulariopsis type, with a clavate conidium which is formed by certain species of *Phyllactinia.*
3. The oidiopsis type, in which the conidiophores are branched and arise from a stoma as in *Leveillula.*

Oidium

Mycelium: Branched, hyaline, septate hyphae.

Conidiophores: Perpendicular to the hyphae, attached to hyphae by cylindrical foot cell

Conidia: Barrel shaped or cylindric, one celled, conidia are produced singly or in chains. Youngest conidium is at the base (basipitally) of the chain and mature are at the tip.

Diseases: Powdery mildew of rubber (*Oidium heveae*), Powdery mildew of *ber (O. erysiphoides* f. sp. *ziziphi)*, Powdery mildew of citrus (*O. citri* and *O. tingitanium*), Powdery mildew of mango (*O. mangiferae*).

Powdery mildew of mango

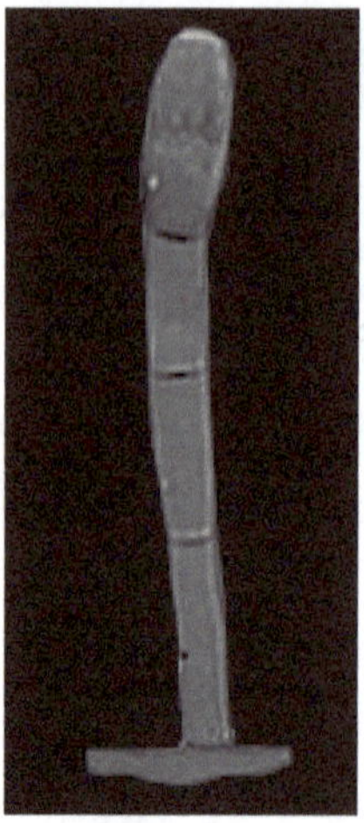

Conidiophore of *Oidium*

Oidiopsis

Mycelium: Endophytic.

Conidiophores: May be branched or unbranched, erect, septate, hyaline and emerge through stomata.

Conidia: Produced singly and cylindrical in shape. Conidia are of two types. a. blunt tip b. pointed tip.

Diseases: Powdery mildew of tomato, eggplant and chilli (*Oidiopsis* sp.), Powdery mildew of *Allium* species (*Oidiopsis taurica*), Powdery mildew of lisianthus, nasturtium, calla lily, impatiens (*O. haplophylli* syn. *O. sicula*).

Other facts

Oidiopsis is the imperfect state of *Leveillula.*

Powdery mildew of chilli

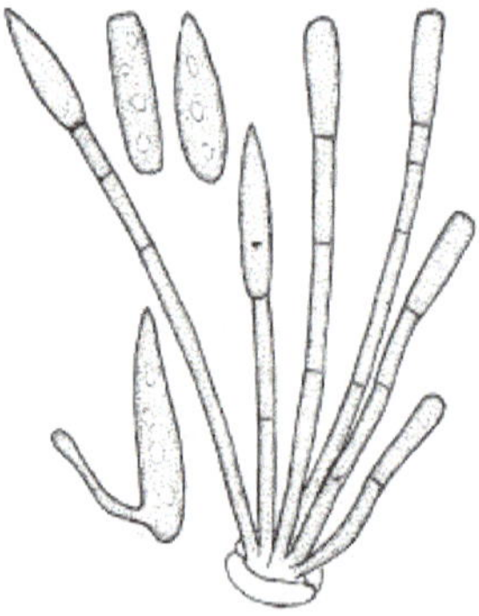

Conidiophore and Conidia of *Oidiopsis*

Ovulariopsis

Mycelium: Partly ectophytic and partly endophytic.

Conidiophore: Hyaline, septate, unbranched and bear a single conidium.

Conidia: Rhomboid in shape. In some species, the conidiphores are spiral in shape.

Other facts

Ovulariopsis is the imperfect state of *Phyllactinia.*

Powdery mildew of Sisam

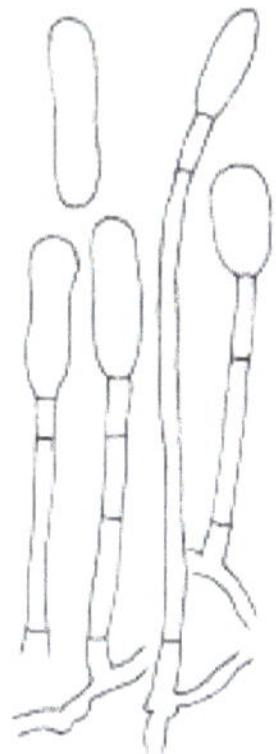

Conidiophore & conidia of *Ovulariopsis*

Erysiphe

Mycelium: Usually superficial.

Conidiophore: Straight or swollen base.

Conidia: In chains.

Cleistothecia: Simple, mycelioid appendages, several asci.

Asci: Cylindrical to oblong, stalked.

Ascospores: Asci contain 2-8 ascospores which are hyaline, oval, elliptical or subcylindric, unicellular.

Diseases: Powdery mildew of cereals and grasses (*Erysiphe graminis*), Powdery mildew of pea and other leguminosae (*E. pisi*), Powdery mildew of cucurbits (*E. cichoracearum*), lettuce (*E. cichoracearum*), sunflower (*E. cichoracearum*), mango (*E. polygoni*), common bean (*E. polygoni*), lucern (*E. pisi*), urd bean (*E. polygoni*), lentil (*E. trifolii*), groundnut (*E. pisi*), coriander (*E. polygoni*), turnip (*E. polygoni*).

Other facts

The imperfect state of *Erysiphe* belongs to form genus *Oidium** (syn. *Acrosporium*).(Gr. Erys=red + siphon=tube)

* the genus is now conserved although invalid as per ICBN.

Powdery mildew of cucurbits

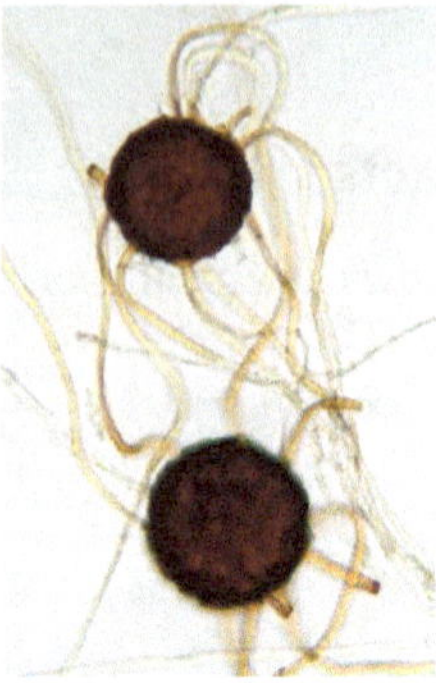

Cleistothecia of *Erysiphe*

Phyllactinia

Mycelium: Mostly superficial as well as partly internal and sends branches through the stomata to the mesophyll tissue where haustoria are formed.

Conidiophores: Produced on superficial hyphae, the base of which is usually straight sometimes twisted.

Conidia: Pointed and are unusual is being formed singly.

Cleistothecia: Bear two types of appendages, radiating unbranched appendages with bulbous bases and a crown of repeatedly branched appendages which secrete mucilage and help in sticking of the fruiting bodies on to twigs and vegetation.

Ascospores: Usually two ascospores in each ascus.

Powdery mildew of Sisam

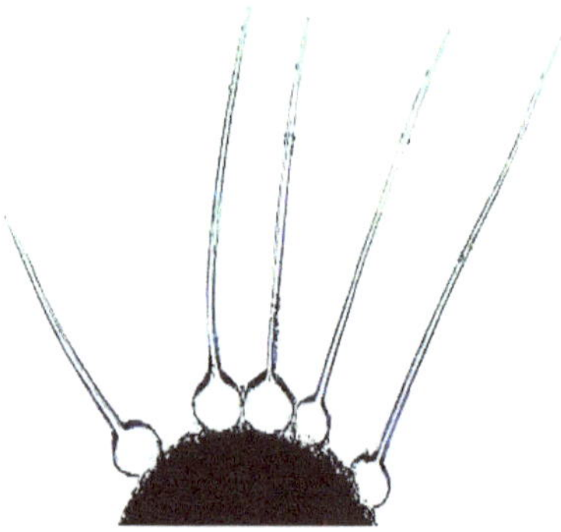

Ascocarp of *Phyllactinia*

Diseases: Powdery mildew of sisam (*P. dalbergiae*), Powdery mildew of hazel (*Phyllactinia guttata*), Powdery mildew of mulberry (*P. guttata*), Powdery mildew of Mulberry (*P. corylea*).

Other facts

The asexual state belongs to the form genus *Ovulariopsis*.

Uncinula

Mycelium: Usually superficial, hyaline, septate, branched,

Conidiophore: Arising perpendicular to the creeping hyphae.

Conidia: Hyaline, oval, in chain, the oldest conidium is at the distal end of the chain.

Cleistothecia: Ascocarp with two or more asci, imperfect state *Oidium* (syn. *Acrosporium*). On cleistothecia the appendages are coiled at the distal end and are brown at the base.

Asci: Each ascocarp contains 4-6, sometimes more, ovate to subglobose ascospores.

Ascospores: Oval to ellipsoid.

Diseases: Powdery mildew of grapes (*Uncinula necator*), Powdery mildew of teak (*U. tectonae*), Powdery mildew of acer (*U. bicornis*).

Other facts

Uncinula is the perfect stage of *Oidium*

The genus is characterized by the production of several asci within each ascoma and appendages with coiled tips (uncinoid) which arise from the peridium.

Powdery mildew of grapes

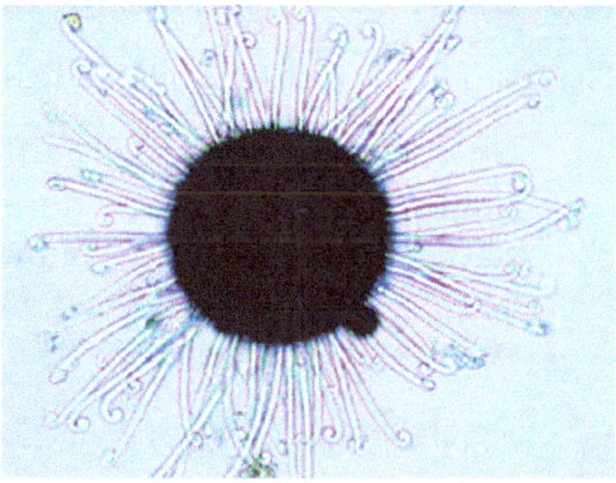

Cleistothecia of *Uncinula*

Podosphaera

Mycelium: Usually superficial, septate, hyaline.

Conidiophore: Arising perpendicular to the creeping hyphae, simple, multisptate and erect, bear chain of conidia.

Conidia: Oval, ellipsoid to barrel shaped, in long chains, the oldest conidium is at the distal end of the chain.

Cleistothecia: Scattered to densely gregarious, appendages are variable in number, myceloid, tortuous, brown. *Podosphaera* is characterized by dichotomously branched appendages and the presence of a single ascus within the ascoma. Ascus contains the ascospores which are ellipsoid to nearly spherical.

Diseases: Powdery mildew of apple and other pome fruits (*Podosphaera leucotricha*), Powdery mildew of hawthorn (*P. clandestina*), plum (*P. clandestina*), apricot (*P. tridactyla*), Powdery mildew of cherry (*P. clandestina*).

Powdery mildew of apple

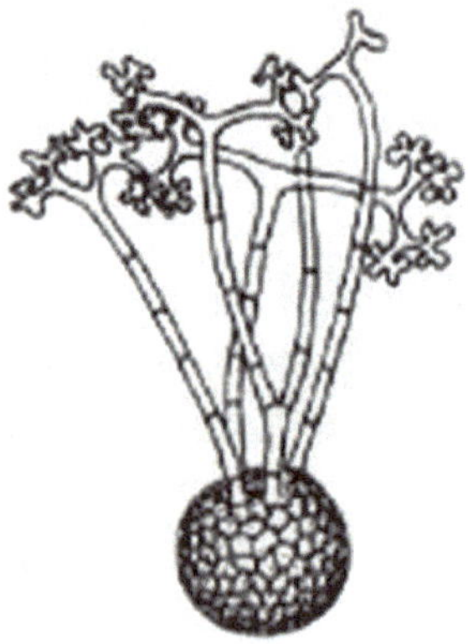

Cleistothecia of *Podosphaera*

11.5. Study of *Puccinia* (different stages), *Uromyces, Hemiliea*

The rust fungi have some unique features.

1. Up to five morphologically different spore states exist in a life cycle of a single species.These structures are generally recognized as basic spore states of the rust fungi as follows: Spermogonia, aecia, uredina, telia and basidia.They are designated often by roman numerals as 0, I, II, III and IV respectively.
2. Heteroecious: Many species need two unrelated groups of host plants to complete their life cycle.

Autoecious: Complete the life cycle on a single kind of host plant.

3. Heteroecious species produce stage 0 and I on one host species and stage II and III on another. Stage IV, the basidial stage is not parasitic.
4. Usually species have narrow and specific host range.

Definitions of spore states according to the ontogenic system (Hiratsuka 1973) are as follows:

Spermatia: Male gametes.

Aeciospores: Non repeating spores produced as a result of dikaryotization, thus in sori normally associated with spermogonia and which give rise to dikaryotic vegetative mycelium.

Urediniospores: Repeating vegetative spores produced on a dikaryotic mycelium.

Teliospores: Basidia producing spores.

Basidiospores: Monokaryotic spores produced on a basidium.

Depending on the number of spores states, three basic types of life cycles are recognized and called macrocyclic, demicyclic and microcylic.

Macrocyclic life cycle has all spore states.

Demicyclic lacks the uredinial states.

Microcyclic lacks both the aecial and uredinial states thus posses only spermogonia and telia.

Spermogonia may be absent from each type but especially the microcyclic life cycle.

Some examples

1. Heteromacrocyclic:

   Puccinia graminis Pers. (black stem rust)
   Cronartium ribicola J. C. Fisch. Ex Rabh. (White pine blister)
   Coleosporium asterum (Diet.) Syd. (Pine needle rust)

2. Automacrocyclic:

   Melampsora lini (Pers.) Lev. (Flax rust)
   Phragmidium mucronatum (Pers.) Schlecht. (Rose rust)
   Uromyces appendiculatus (Pers.) Unger (Common bean rust)

3. Heterodemicyclic:

   Gymnosporangium juniperi-virginianae Schw. (Cedar apple rust).
   Pucciniastrm goeppertianum (Kuehn) Kleb. (Fir needle rust)

4. Autodemicyclic:

   Arthuriomyces peckianus (Howe) Cumm. and Y. Hirat. (Orange rust of Rubus)

5. Microcyclic:

   Puccinia malvacearum Bert. Ex Mont. (Hollyhock rust)
   Coleosporium pinicola (Arth.) Arth. (Pine needle rust)
   Ravenelia pringlei Cumm. (*Accacia* broom rust)

Puccinia striiformis: The survival of the fungus in India is through its uredial stage present on grass hosts at high altitude on the hills. The grass hosts include *Bromus japonicus* found in plains also.

P. recondita: *Isopyrum fumarioides* and species of *Thalictrum* are the alternate hosts for *P. recondita. I. fumarioides* does not occur in India while *Thalictrum* spp. are common in the hills. *Thalictrum* has always been regarded as a non functional host of this rust by all the workers except D'oliveira.

Puccinia

Spermogonia: Subepidermal, flask shaped bodies have an ostiole protruding out through the cuticle. A number of periphysis are present around the ostiole. Inside these pycnia a large number of male reproductive bodies called spermatia (pycnospores) are produced.

Spermatia: Borne singly at the tip of spermatiophores which are formed all along the inside wall of spermogonium. They are minute, thin walled, mostly hyaline and uninucleate.

Receptive hyphae: Cylindrical, long female reproductive structures, form on the upper parts of the spermagonial wall and grow out through the ostiole. **Aecia**: Subepidermal in origin, erumpent, aecidioid with peridium and catenulate, mostly verrucose spores, or uredinoid without peridium and with mostly echinulate spores borne singly on pedicles.

Aeciospores: Usually spherical, smooth or rough walled and binucleate.

Uredinia: Subepidermal in origin, erumpent, without peridium but may have paraphyses, spores borne singly on pedicels, mostly echinulate, pores various.

Uredospores: One celled, subglobose, ovoid or round, light brown or brown in colour, always echinulate.

Telia: Subepidermal in origin, erumpent in most species but remaining covered by the epidermis and sometime divided into locules by stromatic paraphyses in some species.

Teliospores: Typically 2 celled by horizontal septum (but may have 1 celled and sometimes 3 or 4 celled) borne singly on pedicels, smooth or variously sculptured in a few species, basidium external. The telutospore stage in the rust fungus is considered as the perfect stage.

Basidiospore: Mature basidiospore posses a thin wall and lack surface markings. Each spore bears a distinct basal apiculus or hilar region marking the point at which it was attached to its sterigma.The species may be heteroecious or autoecious with a macrocyclic or microcyclic life cycle.

Diseases: Wheat black stem rust (*Puccinia graminnis tritici*), Brown or leaf rust (*P. recondita*), Yellow or stripe rust of wheat (*P. striiformis*), Barley brown or leaf rust (*P. hordei*), Crown rust of oats (*P. coronata*), Rust of bajra (*P. penniseti*), Jowar rust (*P. purpurea*), Maize rust (*P. sorghi, P. polyspora*), Cotton rust (*P. cacabata*), Rust of sugarcane (*P. erianthi* and *P. kuehnii*), Rust of groundnut (*P. arachidis*), Rust of onion (*P. allii*), Rust of mint (*P. menthae*), Rust of guava (*P. psidii*), Rust of sunflower (*P. helianthi*), Rust of chrysanthemum (*P. chrysanthemi*), Rust of safflower (*P. carthami*), Rust of hollyhock (*P. malvacearum*), Rust of antirrhinum (*P. antirrhini*).

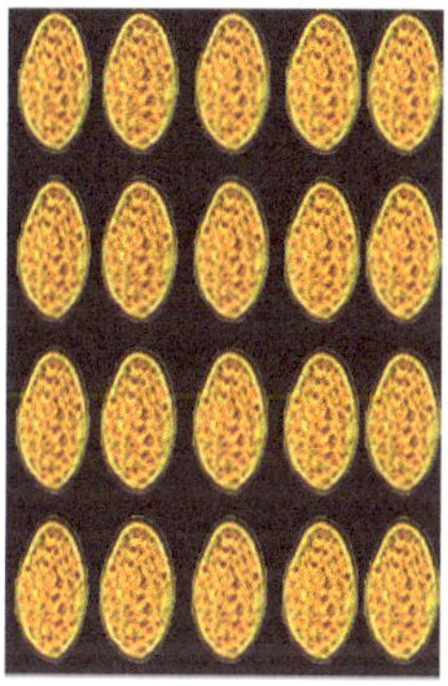
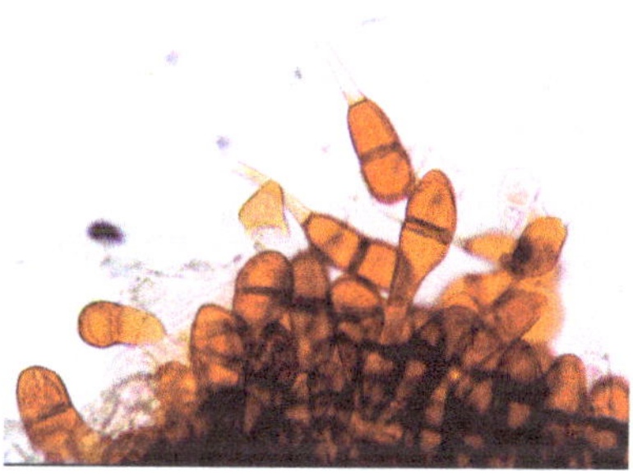

Wheat black stem rust Uredospores and Teliospores of Puccinia

Uromyces

Spermogonia: Subepidermal

Aecia: Subepidermal in origin, erumpent, cupulate.

Aeciospores: borne singly on pedicels, 1 celled, verrucose, elliptical, yellowish brown.

Uredinia: Subepidermal in origin, erumpent.

Uredospore: Spores borne singly on pedicels, globes to oval, usually echinulate, having germ pores, brownish yellow in colour .

Telia: Subepidermal in origin, erumpent or remaining covered by the epidermis.

Teliospores : Borne singly, elliptical to ovate, one celled, stalked, unicellular and papillate with an apical germ pore, basidium external

Diseases: Bean rust (*Uromyces phaseoli typica*), Pea and vicia rust (*U. fabae, U. pisi*), *Trifolium* rust (*U. trifolii*), Rust of *Medicago* (*U. striatus*), Gram rust (*U. ciceris-arietini*), Cowpea rust (*U. vignae*), Rust of foxtail millet (*U. setariae-italicae*), Alfalfa or lucern rust (*U. striatus*), Clover rust (*U. trifolii var. fallens*), Beet rust (*U. betae*), carnation rust (*U. dianthi*).

Other facts

- Uromyces differs from Puccinia only in having 1-celled teliospores.
- Uromyces parasitize monocots and dicots throughout the world.
- Species of Uromyces may be macro or micro cyclic and heteroecious or autoecious.
- Autoecious macrocyclic- *U. appendiculatus, U. fabae.*
- Heteroecious – *U. pisi, U. ciceris-arietini.*

Bean rust

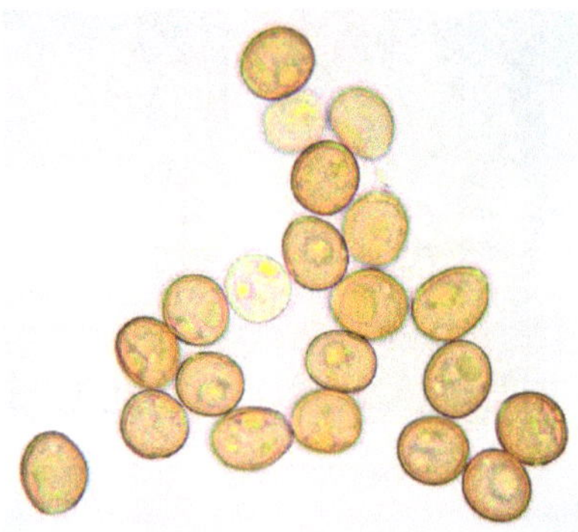

Uredospores of *Uromyces*

Hemileia

Mycelium: Club-shaped with tips bearing numerous pedicels on which clusters of urediniospores are produced.

Uredinia: In sub-stomatal chambers but in some species erumpent through the ruptured epidermis.

Urediniospores: Reniform, with hyaline to pale yellowish wall, borne singly on short pedicels, strongly asymmetrical with 1 flat or concave side, usually smooth and a dorsal convex, strongly warted on the convex side, smooth on the straight or concave side.

Telia: Pale yellowish

Teliospores: Often produced in uredinia, more or less spherical to limoniform, thin walled without germ pores, wall hyaline to yellowish, smooth, unicellular, borne singly on short pedicels usually broader than height, thicker at the apex, pedicel hyaline, often angular, turnip shaped wall uniformly thin, pale, germinate *in situ* without any rest period and are therefore, called leptospores, basidium external.

Diseases: Coffee rust (*Hemileia vastatrix* and *H. coffeicola*).

Other facts

- Species of the genus parasitize both monocot and dicot hosts.
- *Hemileia vastatrix* is an autoecious, microcyclic rust since spermogonia and aecia are lacking.

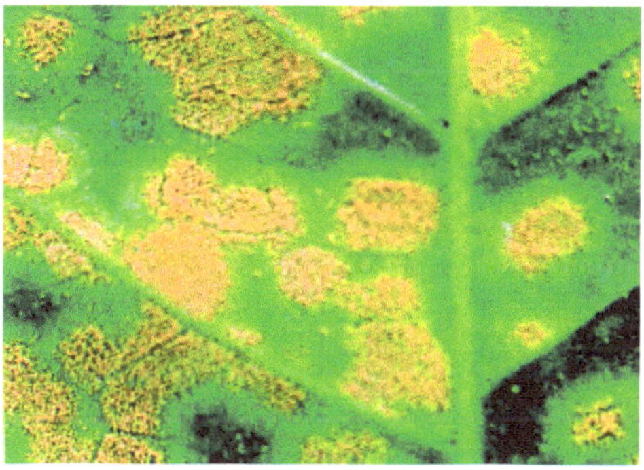

Coffee rust

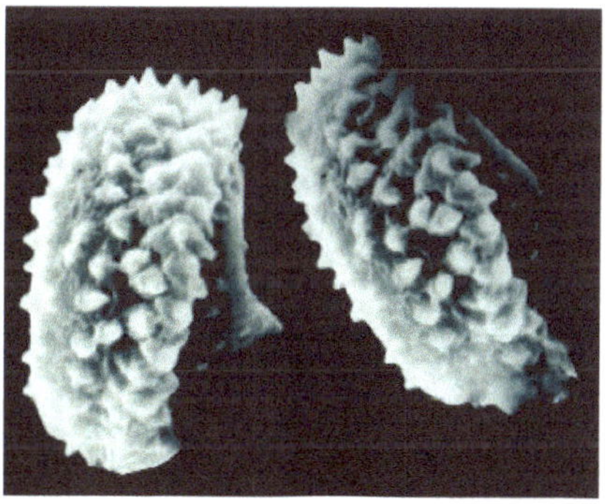

Urediniospores of *Hemileia*

11.6 Study of *Sphacelotheca, Ustilago* and *Tolyposporium*

Sphacelotheca

Sori: Contain 1 celled teliospores, dusty at maturity, but are enveloped in a peridium made up of fungus cells and has a central columella of host tissue.

Columellum: The columellum is made up of host tissues including parenchyma and vascular elements, it is bulbous at the base and sometimes branched at the apex. A dense mass of black to dark brown, smooth, thick walled spores, which are mostly single, fills the space between the columellum and sorus wall.

Spores: The fungus is present in the form of masses of spores enclosed by thin sorus membrane, which easily release the spores. Spores are spherical to elliptical, dark brown with a minutely pitted spore wall. On germination, produces promycelia and sporidia. The sporidia multiply by budding. More recent name used for the *Sphacelotheca sorghi* (Link) Clinton is *Sporisorium sorghi* Ehrenb. Ex Link. In *S. sorghi*- short columella present and spores single.

Sporisorium cruentum (Kuhn) Vanky (Syn.: *Sphacelotheca cruenta* (Kuhn) Potter). Spores have echinulate wall. They germinate to 4 celled promycelium with laterally borne sporidia, the sporidia are unicellular, hyaline, fusiform. At lower temperature sporidia are formed while at higher temperature direct germination is common. *S. cruenta*- long columella present and spores single.

Sporisorium reilianum (Kuhn) Langdon and Fullerton (Syn. *Sphacelotheca reiliana* (Kuhn) Clinton and *Sorosporium reilianum* (Kuhn) Mc Apline). The spores are reddish brown to black, finely echinulate, irregular to spherical. *S. reiliana*- columella absent but a network of vascular tissues present, spores loosely bound into balls.

Diseases: Grain, covered or kernel smut of jowar (*Sphacelotheca sorghi*), Loose smut of jowar (*S. cruenta*), Head smut of jowar (*S. reiliana*), Head smut of maize (*S. reiliana*).

Head smut of maize

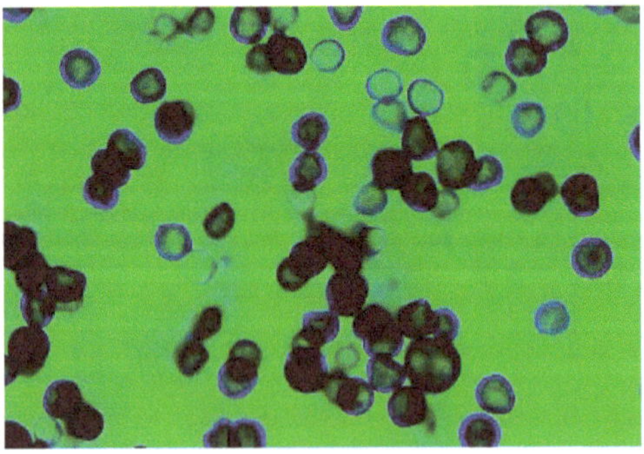

Smut spores of *Sphacelotheca*

Ustilago

Sori: contain 1 celled teliospores, dusty at maturity and are covered by a membrane of host origin.

Spores: The smut spores are round to elliptical, brown in colour. They appear black in mass. The ustilospore have thick, dark brown walls which may be smooth spiny or reticulated. Smut spore, on germination, produces promycelium, into which the diploid nucleus moves. The promycelium may be non septate or if septate, any numbers of septa may be formed. Four or more basidiospores are formed directly on the surface of the promycelium ie. sterigmata are absent, a feature found in the rusts.

Sporidia: Borne on the promycelium near the septa and at the apex. They may be formed laterally or terminally or in both position and are also formed in variable numbers. Sporidia are uninucleate, ovate to oblong, and may form fresh secondary sporidia by budding. Inter sporidial anastomosis are common. A basidiospore on germination, give rise to a monokaryotic primary hypha, which, except in *U. maydis*, exists only for short duration. These hyphae immediately fuse, become dikaryotic and initiate infection.

Chlamydospores: Pale, olive brown, spherical to oval in shape and are decorated with minute echinulations on the wall.

Diseases: Loose smut of wheat (*Ustilago tritici*), Loose smut of barley (*U. nuda*), Loose smut of oats (*U. avenae*), Covered smut of barley (*U. hordei*), Covered smut of oats (*U. kolleri*), Smut of maize (*U. zeae*), Whip smut of sugarcane (*U. scitaminea*), Smut of foxtail millet (*U. crameri*), Smut of barnyard millet (*U. crus-galli*) etc.

Loose smut of wheat

Covered smut of barley

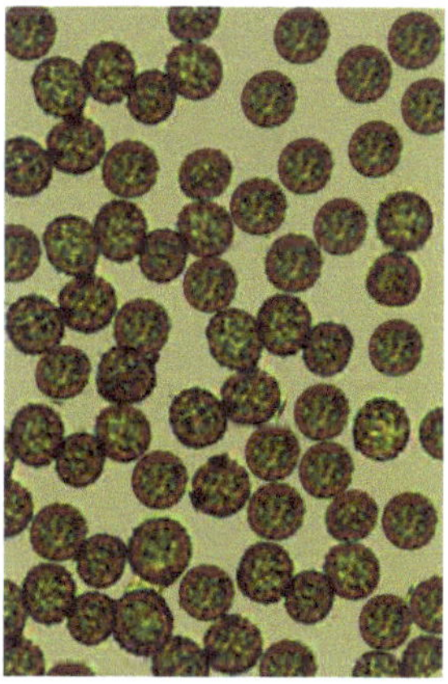

Smut spores of *Ustilago*

Other facts

The genus *Ustilago* (L. ustus= burned) contains about 300 species especially parasitis on members of gramineae.

Tolyposporium

Sori: Consist of permanent spore balls.

Spores: The exposed surface of the spores is covered by flattened echinulations. The individual spores are angular to round, light brown, with roughened spore wall.

Teliospores: The teliospores in the balls are held together. They germinate in water to produce four celled promycelia, bearing the sporidia.

Sporidia: Numerous, hyaline, 1-celled, spindle shaped, single or in chains. They may be formed on the promycelium apically or laterally. *T. ehrenbergii* columella absent but 8-10 vascular strands present.

Diseases: Long smut of Jowar (*Tolyposporium ehrenbergii*), Smut of Bajra (*T. penicillariae*).

Smut of Bajra

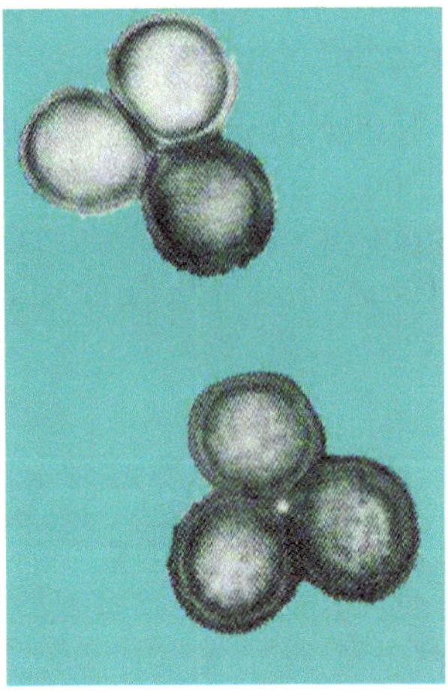

Spores of *Tolyposporium*

11.7 Study of *Agaricus, Pleurotus* and *Ganoderma*

Agaricus

Basidiocarp: White to brown or grey brown cap.

Gills: In a young basidiocarp the free gills may be light in colour- often pink or white but eventually darken chocolate brown in colour. Annulus and stalk present but volva absent. Members of the genus commonly referred as meadow mushrooms, because they typically are found in open, well manured and quite often grassy areas. All the speciec of *Agaricus* are edible except a few such as *Agaricus placomyces, A. xanthodermus* and *A. silvaticus* are poisonous which may cause gastrointestinal disturbance in some individuals.

Agaricus bisporus

Pleurotus

The genus *Pleurotus* contains the so called oyster mushroom. They are spatchulate to kidney shaped, either sessile, lacking a stalk, or have a very short, lateral stalk. The white oyster shell like caps appear on logs or tree stumps in shelf like layer. The stipe is generally eccentric and pileus resupinate in some species.

P. olearium is poisnous, summer species (16-30°C): *Pleurotus flabellatus, P. sapidus, P. sajor-caju* and *P. citrinopileatus*.

Winter species (10-20°C): *P. ostreatus, P. florida, P. eryngii, P. fossulaus* and *P. cornucopiae*.

Pleurotus florida

Ganoderma

Mycelium: Mycelium is the basilar and white villiform.

Fruit bodies: Poroid, either sessile or stipitate, the upper surface of the pileus being shiny as if varnished due to the presence of an amorphous waxy substance secreted by the hyphae.

Ganoderma lucidium

Basidiospores: Ovoid to elliptical, usually truncate at one end and golden brown in colour, with a wall consisting of two layers, the apex at first rounded but later truncated.

Diseases: *Ganoderma lucidium* is a serious root parasite on *Dalbergia, Acacia, Albizzia* etc. *G. pseudoferroum* is parasite on the root of Cacao, coffee, rubber, tea.

Ganoderma lucidium

11.8 Study of *Septoria, Colletotrichum, Pestalotiopsis* and *Pyricularia*

Septoria

Mycelium: Hyaline, septate, branched mycelium which becomes dark with age.

Pycnidia: Sub globose, dark, ostiolated, erumpent, thin walled and brown, immersed in the substratum and are either separate or aggregated but not confluent.

Conidia: Hyaline, smooth, cylindrical, slightly curved, thin walled, 3-5 septate, filiform, continuous or constricted at septa. Germination lateral or terminal.

Conidiophore : Short, conidiogenous cells are hyaline, cylindrical to clavate and aseptate, ampuliform to doliform or lageniform to short cylindrical which occasionally proliferate showing sympodial growth.

Perfect states: *Mycosphaerella* and *Leptosphaeria*.

Diseases: Speckled leaf blotch of wheat (*Septoria tritici*), Foliage blight of celery (*S. apii*), Glume blotch of wheat (*S. nodorum*), Leaf spot of tomato (*S. lycopersici*), Leaf scorch of azalea (*S. azaleae*), Leaf spot of gladiolus (*S. gladioli*), Leaf blotch, stem break and black stem of oats (*S. avenae*), Leaf spot of sweet potato (*S. bataticola*), Spots on canes and leaves of raspberry (*S. rubi*), Speckled blotch of leaf, leaf sheath and glumes of paddy (*S. oryzae*), Leaf spot of passion fruit (*S. passiflorae*).

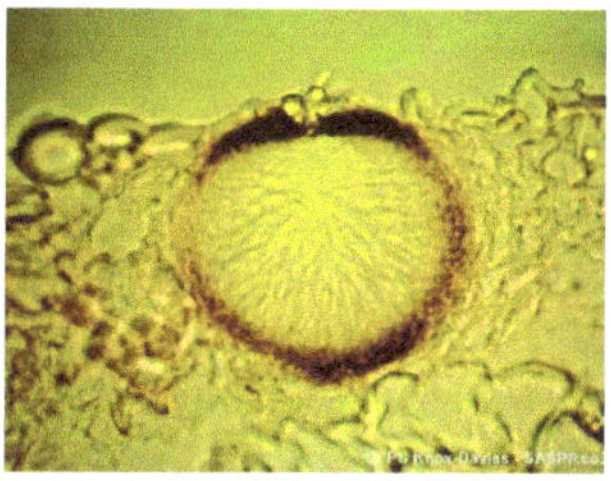

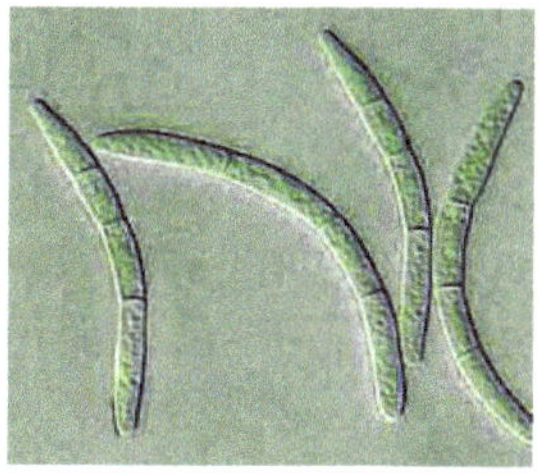

Brown spot of soybean Pycnidia and Conidia of Septoria

Colletotrichum

Mycelium: Hyaline to pale brown, immersed, septate, branched.

Acervuli: Disc shaped or cushion shaped, waxy, may be subcuticular, epidermal or subepidermal and may be either separate or confluent.

Setae: Brown, smooth, straight, unbranched, septate, thick walled and tapered towards the apex. they originate from the pseudoparenchyma at the edge or among the conidiophores.

Conidiophores: Hyaline to brown, short, club shaped, smooth, simple, elongate, septate.

Phialides: Unicellular, hyaline and cylindrical.

Conidia: Hyaline, 1 celled, ovoid or oblong to falcate or lunate (sickle shaped), cylindrical or curved conidia, more or less guttulate, muticate or with the apex prolonged into a simple cellular appendages, typically elongated with pointed to rounded ends, produced from phialids (conidiogenous cells) enteroblastically. Conidia are released in characteristic droplets of mucilage.

Chlamydospore: Produced by *C. falcatum*.

Sclerotia: Sometimes produced in the cultures which are dark brown to black, often confluent, occasionally setose.

Perfect state: *Glomerella*

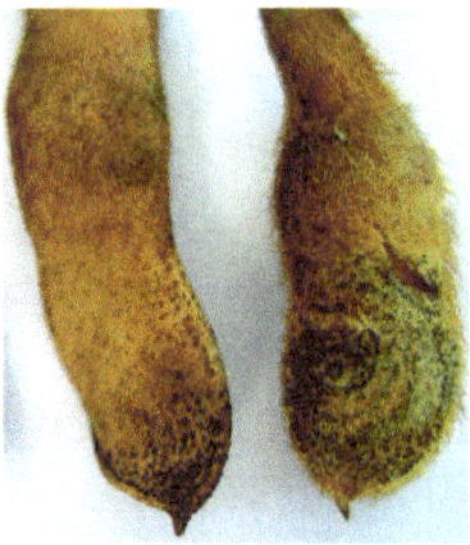

Anthracnose of soybean

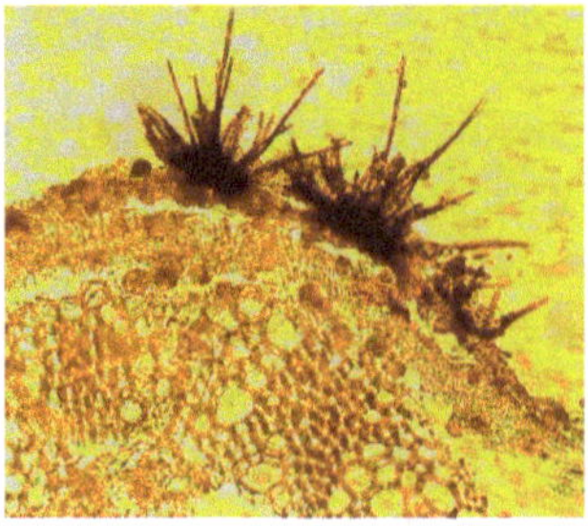

Acervuli

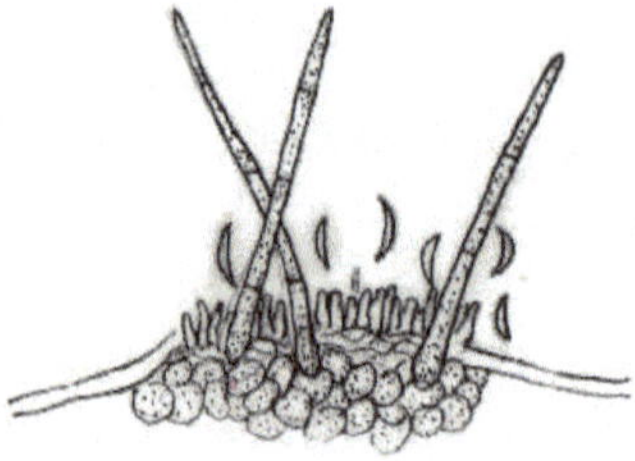

Conidia

Diseases: Anthracnose and pod blight of soybean (*Colletotrichum truncatum*), Red rot of sugarcane (*C. falcatum*), Ripe fruit rot or die back of chillies (*C. capsici*), Ripe rot of grapes (*C. gloeosporioides*), Anthracnose of citrus, mango, papaya, guava, banana (*C. gloeosporioides*), Bitter rot of pome and stone fruits (*C. gloeosporioides*), Coffee berry disease (*C. coffeanum*), Anthracnose of bean (*C. lindemuthianum*), Anthracnose of cucurbits (*C. orbiculare*), Anthracnose of maize (*C. graminicola*), Seedling blight of flax (*C. lini*), Anthracnose of spinach (*C. dematium*), Onion smudge (*C. circinans*), Potato black rot and tomato root rot (*C. coccodes*), Anthracnose of various *Brassica* spp (*C. higginsianum*), Anthracnose of groundnut (*C. truncatum*), Anthracnose of cotton (*C. capsici* and *C. gossypii*), turmeric leaf spot (*C. capsici*), Anthracnose of sorghum (*C. graminicola*).

Other facts: The mucilage has been shown to protect the conidia from desiccation as well as from the toxic effects of polyphenolic compounds released by the host.

Pestalotiopsis

Mycelium: Pink and septate.

Acervuli: Subepidermal and irregularly erumpent through the epidermis or longitudinal cracks may appear.

Conidiophore: Short, branched.

Conidia: Ellipsoid or fusoid and mostly four septae, with 2 or more apical appendages and conidiogenous cell with several proliferation.

Diseases: Gray blight of tea, blight of mango, palms and cotton (*Pestalotiopsis theae*), grey leaf spot, die back, stem canker and petal rot of tea (*P. guepinii*), anthracnose of oil palm (*P. papposa*), leaf spot of groundnut (*Pestalotiopsis* sp).

Other facts:*Pestalotiopsis* (5 celled with acervulus), a segregate of *Pestalotia* (6 celled, pycnidial), contains 45 species.

Grey leaf spot of mango

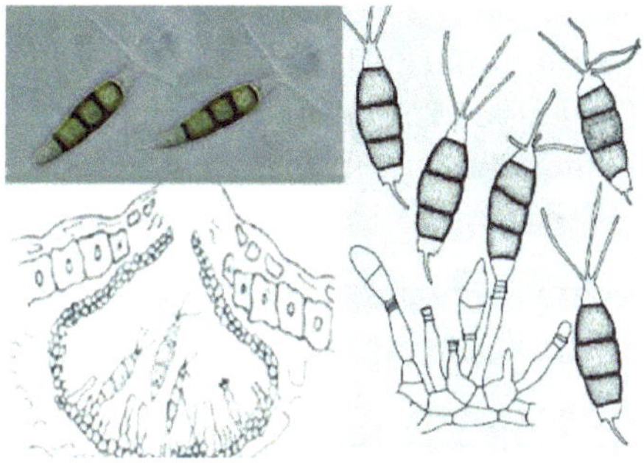

Conidiophore and Conidia of *Pestalotiopsis*

Pyricularia

Mycelium: Hyaline to olvaceous, septate, mostly uninucleate, branched hyphae.

Conidiophore : Long, slender, single or in fascicles, simple or rarely branched. They show sympodial growth and are septate, denticulate and grayish in colour.

Conidia: Conidia Hyaline to pale olive, terminal, narrowly pyriform to obclavate with rounded base, two septate, rarely 1-3 septate. Conidia are borne sympodialy on the conidiophores.

Perfect stage: *Magnaporthe*

Diseases: Blast of paddy (*Pyricularia grisea*), blast of ragi (*P. setariae*), Gray leaf spot of perennial ryegrass and St. Augustine grass **(***Pyricularia* sp.**)**, leaf spot of *Ctenathe* (*P. oryzae*), leaf spot of safflower **(***Pyricularia sp.*), leaf spot of millet (*P. grisea*), oval leaf spot of graminea, pitting disease, ryegrass blast, and Johnson spot (*P. grisea*).

Blast of paddy

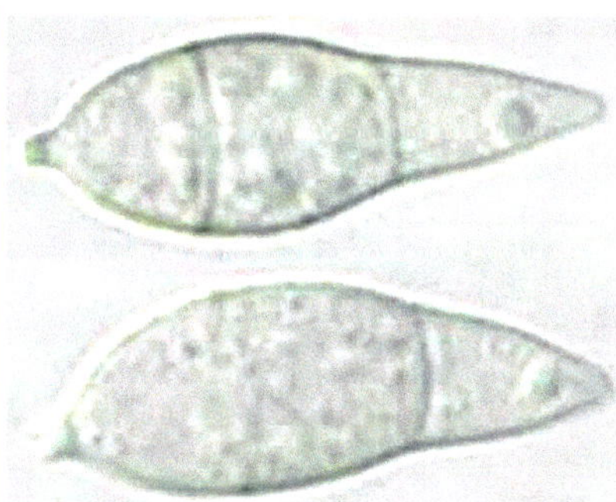

Conidia of *Pyricularia*

11.9 Study of *Aspergillus, Penicillium, Trichoderma* and *Fusarium*

Aspergillus

Conidiophore: Upright, simple, usually arise singly from the somatic hyphae. The long and erect conidiophores arise from a particular thick walled cell known as the foot cell. It forms a swollen vesicle on which bottle shaped structures called sterigmata or phialides or conidiogenous cells arise.

Conidia: Phialospores, one celled, globose, often variously coloured in mass, as the phialides reach maturity, they begin to produce chains of conidia at their tips in basipetal succession with the youngest conidium at the base and the oldest at the tip. The fungus produces cleistothecia in which asci and ascospores are produced.

Perfect state: *Eurotium, Sartorya, Emricella* and *Hemicarpentels*.

Diseases: Smut of fig, black mold of cotton bolls, ear rot of corn, black mold of onions (*Aspergillus niger*).

Other facts

- The genus *Aspergillus* monographed by Thom and Church (1926), Thom and Raper (1945) and Raper and Fennell (1965).
- The conidiophore of *Aspergillus* somewhat resembles the aspergillum used in roman Catholic Church and is named after it.
- *A. flavus* is responsible for the production of aflatoxin, which can induce liver cancer in man and poultry. Members of the form genus *Aspergillus* cause a group of diseases known as aspergillosis.
- *A. niger* produces citric and glucanic acid, *A. flavus* produces kojic acid.
- *A. niger* is widely used in the bioassay of available amounts of copper, magnesium, potassium and molybdenum in the soil.

Conidiophore and Conidia of *Aspergillus*

Black mold of onion

Penicillium

Mycelium: Branched, septate, usually colorless.

Conidiophores: Arising from the mycelium singly or less often in synnemata, branched near the apex, ending in a group of phialides, resembling a brush like head (penicillus). *Penicillium* species is based chiefly on the mono or biverticillate nature of the conidiophore.

Conidia: Hyaline or brightly coloured in mass, 1-celled, mostly globose or ovoid, in dry basipetal chain. Conidia are held together in chains by connectives, phialospores.

The sexual reproduction is through the formation of characteristic cleistothecia containing asci and ascospores.

Some species have perfect states: *Talaromyces, Eupenicillium* and *Carpenteles*.

Diseases: Blue mold of apple (*Penicillium expansum*), green mold of citrus (*P. digitatum*), blue mold of citrus (*P. italicum*), storage rot of Gladiolus (*P. gladioli*).

Other facts

- The genus was first described in the scientific literature by Johann Heinrich Friedrich Link in his 1809 work *Observationes in ordines plantarum naturales*.
- The genus has been monographed by Raper and Thom (1949), Pitt (1979/80) and Ramirej
- The genus name is derived from the Latin word *penicillium*, meaning "painter's brush" (L. Penicillum = small brush).

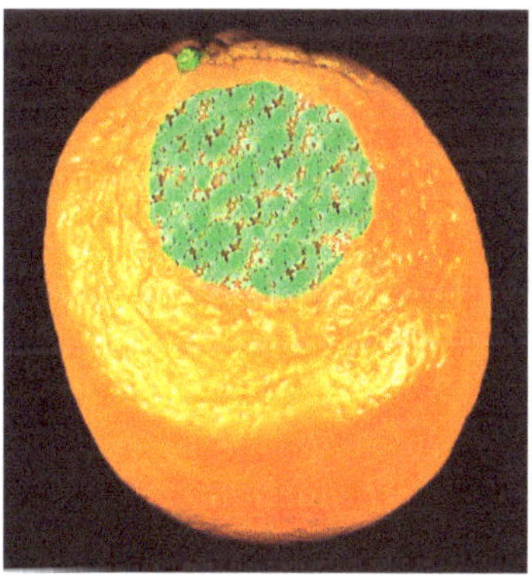

Blue mold of citrus

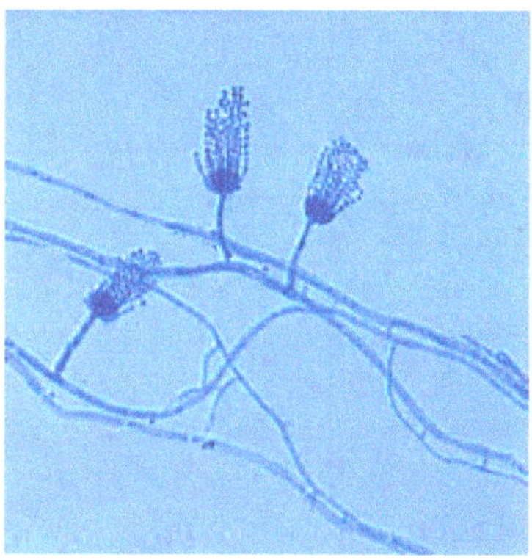

Conidiophores and Conidia of *Penicillium*

- *P. roquefortii* and *P. camembertii* are used for the manufacture of blue cheeses and camembert cheese.
- *P. notatum* and *P. chrysogenum* are the producer of the life saving drug, the penicillin which is active against Gram positive bacteria. *P. chrysogenum* used commercially.
- *P. griseofulvum* is used in the production of Griseofulvin, it is an antifungal drug and a potential chemotherapeutic agent.
- *Penicillium* used in the production of fermented foods, as a flavoring agent.
- *P. chrysogenum*, *P. steckii*, *P. notatum*, *P. cyclopium*, and *P. nalgiovensis* affect fuels; *P. chrysogenum*, *P. rubrum*, and *P. verrucosum* cause damage to oils and lubricants; *P. regulosum* damages optical and protective glass.

Trichoderma

Conidiophores: Hyaline, much branched, not verticillate, phialides single or in groups, producing hyaline or green, unicellular phialospores, gathering in balls at the mouths of the phialides.

Conidia: Hyaline, 1-celled, ellipsoidal in most of species, phialospores, borne in small terminal clusters, usually recognized by its rapid growth and green patches or cushions of conidia, saprophytic on soil or on wood.

Chlamydospores: Typically unicellular, subglobose, but in some species Chlamydospores are multicellular (e.g. *T. stromaticum*).

Some Species: *Trichoderma viride, T. harzianum, T. hamatum, T. koningii, T. reesei, T. ovalisporum, T. atroviride, T. aureoviride, T. austrokoningii, T. brevicompactum, T. candidum, T. cinnamomeum, T. citrinoviride, T. crassum, T. dorotheae, T. effusum, T. estonicum, T. fertile, T. gelatinosus, T. helicum, T. intricatum, T. longibrachiatum, T. longipile, T. minutisporum, T. polysporum, T. pseudokoningii, T. pubescens, T. rossicum, T. saturnisporum, T. sinensis, T. strigosum, T. taiwanense, T. thailandicum, T. tomentosum, T. velutinum, T. virens.*

Teleomorph: *Hypocrea*

Diseases: Blotch of white button mushroom (*Trichoderma viride, T. hamatum* and *T. harzianum*), *T. aggressivum* (formerly *T. harzianum* biotype 4) is the causal agent of green mold, a disease of cultivated button mushrooms.

Green mold of cultivated button mushroom

Conidiophore and Conidia of *Trichoderma*

Other facts

- *T. reesei* is used to produce cellulase and hemicellulase
- *T. harzianum* is used to produce chitinase.
- *T. viride* is the causal agent of green mold rot of onion.

Fusarium

Mycelium: Septate, branched, intracellular

Conidiophores: Variable, slender and simple or stout, short, branched irregularly or bearing a whorl of phialides, single or grouped into sporodochia.

Conidia: Hyaline, variable, principally of two kinds.

Macroconidia- hyaline, long, multiseptate, produced on conidiophores which may be solitary and simple or aggregated, slightly curved or bent at the pointed ends, typically (boat) canoe-shaped.

Microconidia- Ovoid to short cylindric, non septate to one septate, some conidia intermediate, 2 or 3 celled, borne singly or in chains or more commonly in spore balls.

Chlamydospore: Spherical to oval, thick walled, single or in chains of two or three, terminal or intercalary. In addition some species produce sclerotia.

Perfect state: *Gibberella, Nectria, Calonectria* and *Micronectriella*.

Diseases: Panama wilt of banana (*Fusarium oxysporum* f. sp. *cubense*), baknae disease of rice, maize (*F. moniliforme*), pigeon pea wilt (*F. udum*), pea wilt (*F. oxysporum* f. sp. *pisi* and *F. solani* f.sp. *pisi*), chick pea wilt

(*F. oxysporum* f. sp. *ciceris*), flax or linseed wilt (*F. oxysporum* f. sp. *lini*), tomato wilt (*F. oxysporum* f. sp. *lycopersici*), cotton wilt (*F. oxysporum* f. sp. *vasinfectum*), cucumber wilt (*F. oxysporum* f. sp. *cucumerinum*), cereal root rot (*F. avenaceum* and *F. culmorum*), Narcissus basal rot (*F. bulbigenum*), Dry rot of potato tubers (*F. caeruleum*), Dry rot of bean (*F. solani f. sp. phaseoli*), Cabbage yellows (*F. conglutinans*), Coriander wilt (*F. oxysporum* f. sp. *coriandri*), cumin wilt (*F. oxysporum* f. sp.*cumini)*, watermelons wilt (*F. oxysporum* f. sp. *niveum*), Ginger yellows (*F. oxysporum* f. sp. *zingiberi*), Wilt of Guava (*F. oxysporum* f. sp. *psidii*).

Chickpea wilt

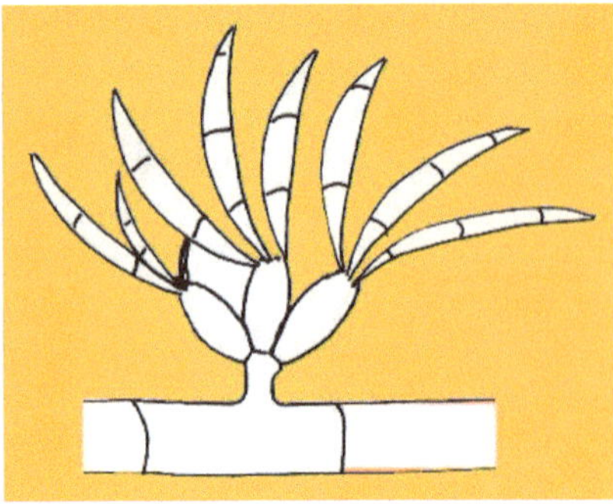

Conidiophore and Conidia of *Fusarium*

Other facts

The genus was erected by Link (1809)

11.10 Study of *Helminthosporium, Drechslera, Alternaria, Stemphylium, Cercospora, Phaeoisariopsis, Rhizoctonia* and *Sclerotium*

Helminthosporium

Mycelium: Mycelium dark often in substrate, sub hyaline to light yellow, olive brown.

Conidiophores: single or clustered, tall, erect, brown to dark brown, fasciculate, determinate, septate, darker at the base than at the top, bear conidia in a synchronous manner.

Conidia: Large transversely septate, often appearing in whorls, single, subhyaline to brown, obclavate, phragmosporous, pseudoseptate with prominent basal scar porospores, produced apically and laterally from the determinate conidiophore, terminal conidium youngest.

Perfect state: *Cochliobolous*

Disease: Silver scurf disease of potato (*Helminthosporium solani*)

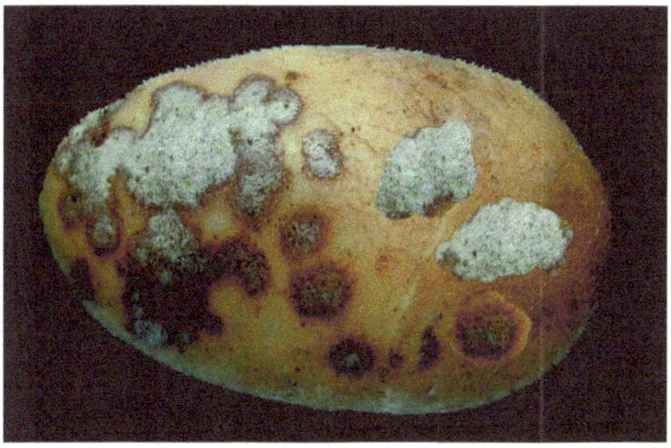

Silver scurf of potato

Conidia of *Helminthosporium*

Other facts

Shoemaker restricts the genus *Helminthosporium* to lignicolous species and classifies graminicolous in *Bipolaris* and *Drechslera*.

Drechslera

Mycelium: Subhyaline to light yellow, intercellular.

Conidiophore: Septate, brown, olive black conidiophores, mostly simple, either singly or in groups.

Conidia: Elliptical, obclavate or clavate, dark, provided with minute hilum, porospores, several celled (phragmosporous), singly at apex through small pores in acropetal succession, continuing growth sympodially from a point below apex and then forming a second spore on new apex, germinating from any or all cells, parasitic or saprophytic.

Perfect states: *Cochliobolous, Pyrenophora, Pleospora* and *Trichometasphaeria.*

Diseases: Brown leaf spot of rice (*D. oryzae*), eye spot and seedling blight of sugarcane (*D. sacchari*), bird eye spot of rubber (*D. heveae*), seedling blight, foot rot and ear blight of barley (*D. teres*), leaf stripe of barley (*D. graminca*), northern leaf blight of maize (*D. turcica*), brown stripe of sugarcane (*D. stenospila*), minor leaf spot disease on coconut, oil palm and sugarcane (*D. halodes*), leaf spot of various grasses, cereals and millets (*D. rostrata*), seedling blight and foot rot of ragi (*D. nodulosus*), leaf blotch of oat (*D. avenae*), leaf spot of sorghum (*D. sativum, D. halodes, D. rostratum*), leaf blight of sorghum (*D. turcicum).*

Brown spot of rice

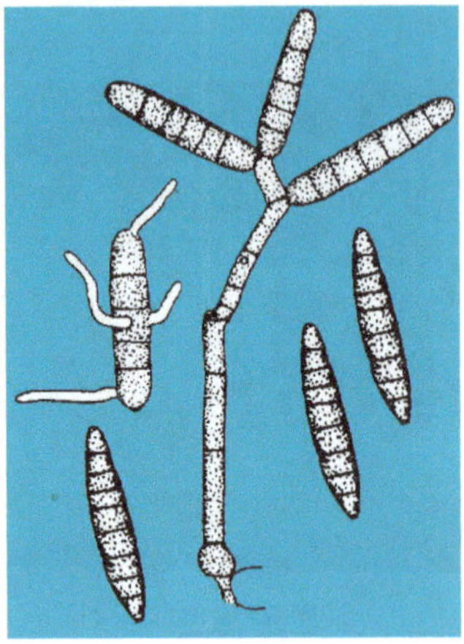

Conidiophore and Conidia of *Drechslera*

Other facts

- The genus was erected by Ito (1930) for those species of *Helminthosporium* with cylindrical conidia that germinate from any cell.
- In *Drechslera* all the cells of conidium are capable of germination.
- While in *Bipolaris* conidia germinate only by the two end cells, Southern corn blight (*Bipolaris maydis,* also known as *Cochlibolus heterostrophus* as its teleomorph state).
- In both *Drechslera* and *Bipolaris* the conidiophores are indeterminate, extending by sympodial growth and are thus distinct from *Helminthosporium*, in which the production of the apical conidia terminates the growth of the conidiophores.

Alternaria

Mycelium: Septate, branched, hyaline, light brown which become darker with age.

Conidiophore: Dark, mostly simple, singly or in groups, erect, straight or curved, determinate or sympodial, rather short or elongate, more or less cylindrical but often slightly swollen at the base, smooth, often geniculate.

Conidia: Porospore, dark, typically with both cross and longitudinal septa (muriform), variously shaped, obclavate to elliptical or ovoid, beaked, frequently borne acropetally in simple or branched acropetal chain but in some instances they are borne single as in *A. solani*.

Perfect state *Pleospora*

Diseases: Early blight of potato and tomato (*Alternaria solani*), leaf blight of wheat (*A. triticina*), Stackburn of rice (*A. padwickii*), leaf spot of crucifers (*A. brassicicola*), purple blotch of onion (*A. porii*), cucurbit leaf spot (*A. cucumerina*), leaf spot of tobacco (*A. alternata*), leaf spot and pod necrosis of soybean, pod and seed decay in soybean (*A. tenuissima*).

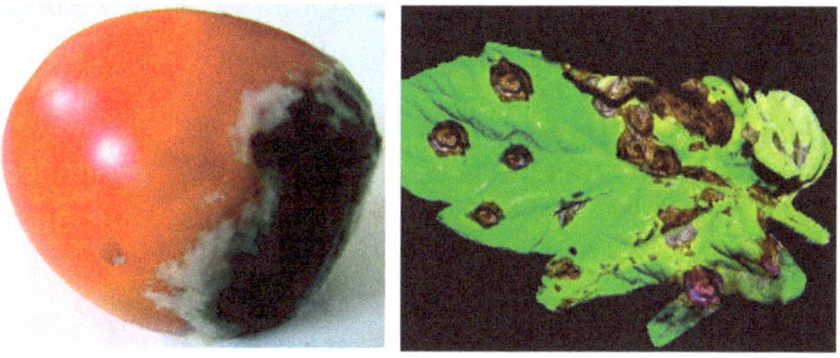

Early blight of tomato

Conidia of *Alternaria*

Other facts

- Existence of chlaymydospores (*A. solani, A. porri*) in the fungus is also reported.
- Fungus also produces spherical black Sclerotia (*A. padwickii*).

Stemphylium

Conidiophore: Dark, mostly simple with darker terminal swelling, short or long, bearing a single, terminal conidium or successive conidia on new growing tips, conidiophore often proliferating through old conidial scar.

Conidia: Porospore, dark, with cross and longitudinal septa, variable in shape, frequently globose, broadly ellipsoid or ovoid, often constricted at major septum, parasitic or saprophytic.

Leaf Blight of Onion

Conidiophore and Conidia of *Stemphylium*

Disease: Floret sterility in wheat (*Stemphylium tritici*), Stemphylium leaf blight of Onion and garlic (*S. vesicarium*).

Cercospora

Mycelium: Septate, light brown coloured.

Conidiophores: Darker at the base and lighter towards the tip, simple, arising in clusters, mostly once geniculate, unbranched, septate, sympodially extending, cicatrized (ie. with conspicuous scars) and bursting out of leaf tissue.

Conidia: Long, filiform (scolecospores), several celled, hyaline or pigmented conidia borne in acropetal succession. Base of a conidium is usually rounded and bears circular scars marking the attachment to the conidiophore. Stromata are dark brown.

Perfect state: *Mycosphaerella*

Diseases: Leaf spot or tikka disease (*Cercospora personatum* and *C. arachidicola*), leaf spot of pigeon pea (*C. indica*), leaf spot of cucurbits (*Cercospora* spp.), frog eye leaf spot of tobacco (*C. nicotianae*), leaf spot of castor (*C. ricinella*), leaf spot of sesamum (*C. sesami*), leaf spot of sorghum (*C. sorghi*), leaf spot of tapioca (*C. henningsii*), leaf spot of sugarbeet (*C. beticola*), sigatoka of banana (*C. musae*).

Other facts

- The genus was first monographed by Chupp (1954). Vasudeva (1963) published a monograph on Indian Cercosporae. Katsuki (1965) wrote a monograph on *Cercospora* spp of Japan.
- *C. apii*, the cause of leaf spot of celery and responsible for causing face lesions in human beings, disease is known as cercosporomycosis. It is based on single but thoroughly studied case in Indonasia.

Sigatoka of banana

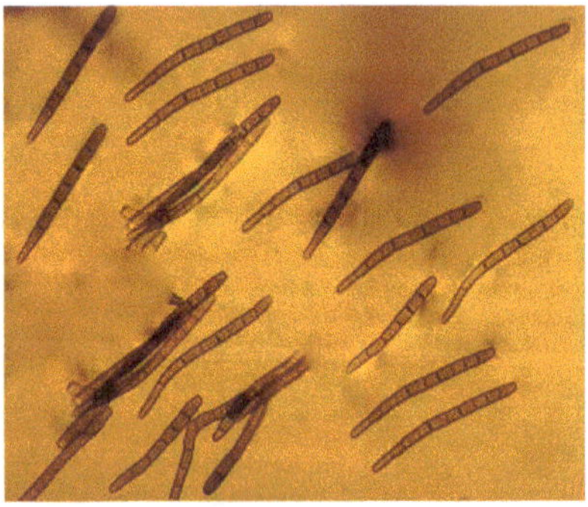

Conidia of *Cercospora*

Phaeoisariopsis

Stroma: Dense, pseudoparenchymatous, up to 130 μm in diameter.

Conidiophores: Numerous, pale to olivaceous brown, smooth, 1-3 geniculate, conidial scars conspicuous, prominent.

Conidia: Medium olivaceous, cylindrical, obclavate, usually straight or slightly curved, wall usually finely roughened, rounded at the apex, base shortly tapered with a conspicuous hilum, 1-9 septa not constricted, mostly 3-4 septate.

Perfect state: *Mycosphaerella*

Diseases: Late leaf spot of groundnut (*Phaeoisariopsis personata*)

Other facts

- Pathogen soil borne
- The nomenclature of this fungus has undergone several changes in the literature. Until recently the combination, *Cercosporidium personatum* (Berk. & Curt.) Deighton was widely used.
- In 1983, J.A. Von Arx reorganized the anamorphs of the genus *Mycosphaerella*.

Late leaf spot of groundnut

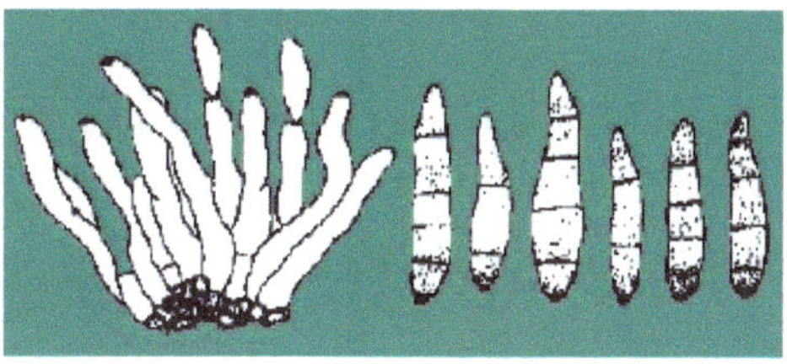

Conidiophores and conidia

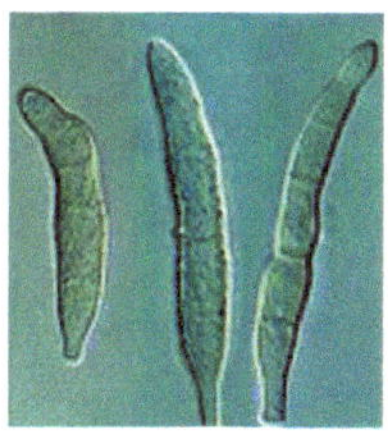

Conidia of *Phaeoisariopsis*

Rhizoctonia solani

Asexual fruit bodies and conidia lacking.

Mycelium: Hyaline to deep brown, 4–15 ìm wide and tend to branch at right angles. Lateral branches were constricted at their point of origin and a septum occurs in the branch near its junction. A special type of cross wall within the hyphae, called a dolipore septum is present. No asexual spores are formed by the mycelium. Fungus produces brown-to-black sclerotia, uniform in texture and highly variable in shape and size.

Perfect state: The teleomorph of *R. solani* is *Thanatephorus cucumeris*. It forms club-shaped basidia. Four spores are produced on each basidium, with four apical sterigmata on which oval, hyaline basidiospores are borne. Each basidiospore has a single nucleus.

Web blight of mungbean

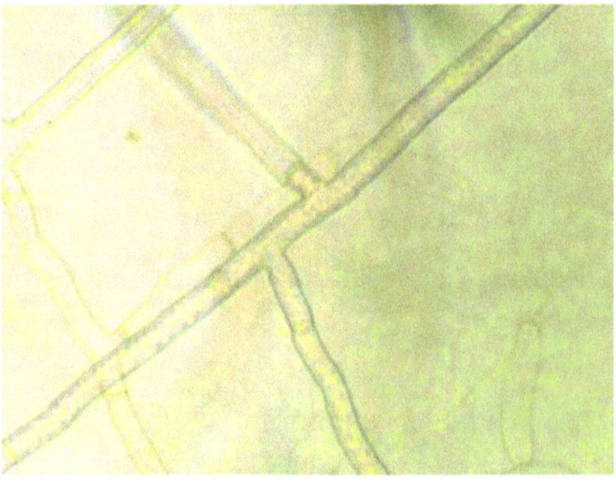

Mycelium of *Rhizoctonia*

Diseases: Root rot of soybean, root rot and damping off disease of fenugreek, sheath blight of rice, stem canker and black scurf of potato, bottom rot and head rot of cabbage, aerial blight or web blight of soybean, web blight of mung bean.

Other facts

- *R. solani* is subdivided into anastomosis groups (AG) based on hyphal fusion between compatible strains.
- In nature, *R. solani* reproduces asexually and exists primarily as vegetative mycelium and/or sclerotia.

Sclerotium rolfsii

Asexual fruit bodies and conidia lacking.

Mycelium: The hyphae of *S. rolfsii* are well branched, compact, hyaline, sparsely septate, apparently thin-walled and often grow in the form of strands. Clamp connections are common. A unique and prominent feature of *S. rolfsii* is the presence of concentric and parallel lamellar structures in the hyphal cells. The hyphae are irregularly partitioned by septa associated with dolipore structure. The "septal pore cap" or "parenthesome" is porous and lamellar.

Sclerotia: Sclerotia forming on a host tend to have a smooth texture. They are globose or irregular, initially white, but sclerotia quickly melanize to a dark brown coloration. Townsend and Willetts recognize four zones in the mature sclerotium:

i) thick skin, ii) rind of thickened cells, iii) cortex of thin walled cells, and iv) medulla containing filamentous hyphae.

Perfect state: *Athelia rolfsii* (Curzi) Tu and Kimbrough (basidial stage)

Diseases: Southern blight (damping-off and stem rot) of soybean, Stem rot of groundnut, Root-rot of sugarbeet, Southern blight of pepper, Southern blight on tomato, Wilt of potato, Sclerotial blight and Circular spot of sweet potato, Collar rot of pumpkin, Southern blight and fruit decay of cucumber, Muskmelon and watermelon, Southern blight of corn, Seedling blight and foot and root rot of wheat, root rot and collar rot of lentil, Seedling blight in rice, Collar rot of betelvine, Leaf and sheath blight of sorghum, Red rot of leaf sheath and sprout rot, Collar rot of tobacco, Wilt and collar rot of sunflower, Bulb rot of onion, Stem rot of garlic, Collar rot of brinjal, Collar rot of chickpea, Wilt of zinnia, Stem rot of chrysanthemum, Corm rot of gladiolus.

Plant infected with *Sclerotium*

Southern blight of corn

Other facts

- The name *Sclerotium* was first introduced by Tode (1790).
- *Sclerotium rolfsii* Sacc. was first reported by Rolfs (1892) as a cause of tomato blight in Florida. Later, Saccardo (1911) named the fungus as *S. rolfsii*.
- Many *Sclerotium* species do not or rarely reproduce sexually and are known only from their asexual stage.
- *Sclerotium rolfsii* produces oxalic acid (Phytotoxin).

Glossary

Acervuli	A mat of hyphae giving rise to short conidiophores closely packed together forming a bedlike mass that becomes erumpent on the tissue of a host plant.
Acropetal	A chain of spores with the youngest at the tip or distal end of the chain.
Aecidioid	Aecidium like.
Anamorph	An asexual state of a fungus.
Apiculus or hilar (of a spore)	A short projection at one end; a projection by which it was fixed to sterigma, apicule, hilar appendage.
Aplerotic	Condition of oospore, when it does not completely fill the oogonium.
Appressorium	A swelling on a germ-tube or hypha, especially for attachment in an early stage of infection.
Ascoma (pl. ascomata)	An ascus containing structure, ascocarp.
Autoecious	A parasitic fungus that completes its entire life cycle on the same host.
Basipetal	A chain of spores with the youngest at the base or proximal end of the chain.
Catenate	Produced in a chain.
Catenulate	Conidia formed in chains of two or more. eg. Cladosporium

Chlamydospores	A thick walled asexual spore formed by the modification of a cell of a fungus hypha, that generally functions as a resting spore.
Cicatrized	Scar formation at the site of a healing wound.
Circinate	Recurved.
Cleistothecium	A completely closed ascocarp.
Coenocytic	A multinucleate, aseptate thallus.
Collumellate	With columella.
Columella	A sterile structure within a sporangium or other fructification, often an extension of the stalk.
Confluent	Coming together
Conidiogenous cell	A cell that forms conidia. e.g., a phialide.
Cupulate	Cup shaped, deeper than saucer shaped.
Diclinous	Having the oogonium and the antheridium on different hyphae.
Dichotomous	A type of hyphal branching into two equal forks.
Eccentric	Not precisely circular.
Echinulate	Having spines or other sharp projections from cell wall.
Enteroblastically	When inner wall is involved in conidia formation.
Erumpent	Breaking out through the surface of the substratum.
Filiform	Long, needlelike.
Fusiform	Spindlelike, narrowing towards the ends.
Geniculate	Bent like a knee.
Guttulate	Having one or more oill ike drops inside.
Heteroecious	Requiring two different kinds of plants (hosts) to complete its life cycle (particularly in rust fungi).
Hyaline	Colourless.
Lageniform	Flask-shaped.
Limoniform	Shaped like a lemon.

Monoclinous	Having the antheridium on the oogonial stalk.
Muriform	Having both transverse and longitudinal septa.
Muticate	Not sharpe at the end.
Mycelioid	Mycelium like.
Mycelium (pl. mycelia)	The mass of hyphae making up the thallus of a fungus.
Ontogenic	Based on development (mostly in context with conidial).
Ostiole	An opening or pore in an ascocarp or a pycnidium.
Paraphyses	Sterile, basally attached structures in the hymenium of perithecia and apothecia.
Penicillate	A brushlike cluster of sporogenous cells on a conidiophore as in Penicillium.
Peridium	The outside covering or a wall of a fructification.
Periphysis	Short, hairlike growths in the form of a fringe lining the inside of an ostiole or of a pore in a stroma.
Phialides	End cell of a conidiophore of fixed length with one or more open ends through which a basipetal succession of conidia develop.
Phialospores	Conidium borne by a phialide without an increase in length of the phialide itself.
Phragmospore	Spore having two or more transverse septa.
Poroid	Having pores.
Porospore	Spore developed by expansion through an opening of fixed dimension.
Reniform	Kidney shaped.
Resupinate	Lying flat on the sub-stratum with the hymenium on the free surface.
Scolecospores	An elongated, needlelike or wormlike spore.
Setose	With setae.
Sorus	A mass of sporangia or spores.
Spermogonium (pycnium)	A fruiting body of rust fungi in which spermatia (male gametes) are formed.

Sporangiophores	A specialized hypha that bears a sporangium.
Sporangium	A container or case of asexual spores. In some cases it functions as a single spore.
Sporidium	The basidiospore of a smut fungus.
Sporodochia	A cushion shaped structure covered with conidiophores.
Sterigma (pl. Sterigmata)	A small hyphal branch or structure, which supports a sporangium, a conidium, or a basidiospore.
Stolons	Connecting hyphae between two sporangiophores (Rhizopus).
Stroma	A compact somatic structure on which or in which fructifications usually are formed.
Sympodial	Growth or branching of a conidiophore or sporogenous cell arising beneath or behind the previous conidium and pushing it to one side.
Synnema (pl.Synnemata)	A group of conidiophores cemented together and forming an elongated spore-bearing structure.
Urediniospores	A reddish spore that is produced in the uredinium of a rust fungus
Ustilospore	Teliospores of smut fungi.
Vasculum	A vasculum is a container used by botanists to keep field samples viable by maintaining a cool, humid environment.
Verrucose	Wart like.
Verticillate	Having a whorl of three or more branches or sporogenous cells arising at the same level.
Zygospores	The sexual or resting spore of zygomycetes produced by fusion of two gametangia.

FSC
www.fsc.org
MIX
Papier | Fördert gute Waldnutzung
FSC® C083411

Zeitfracht Medien GmbH
Ferdinand-Jühlke-Straße 7
99095 Erfurt, Deutschland
produktsicherheit@kolibri360.de